Nontechnical Guide to DEEPWATER DRILLSHIPS

PETER TOMDIO

THIS BOOK IS DEDICATED TO
MY WIFE LIZ AND
OUR THREE CHILDREN

Library of Congress Control Number: 2023941131

Edited by Ian Robertson
Designed by Jack Chappell
Cover design by Chris Bower
Type set in Univers

ISBN: 978-0-7643-6662-8
Printed in China

Published by Schiffer Publishing, Ltd.
4880 Lower Valley Road
Atglen, PA 19310
Phone: (610) 593-1777; Fax: (610) 593-2002
Email: Info@schifferbooks.com
Web: www.schifferbooks.com

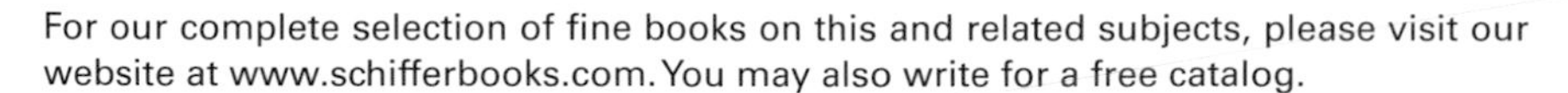

CONTENTS

FOREWORD

The intent of this book is by no means to make the reader an expert in deepwater drillship operations, but rather to help familiarize the experienced mariner with what makes these vessels drilling units; the experienced driller or tool pusher who has worked on land rigs and wants to know what makes these drilling units vessels. To the shore-based support staff, young engineering graduates, new mariners venturing into deepwater-drilling operations, or analysts who cover these vessels but have never had the opportunity to step on these vessels, this is your gateway. In the merchant marine, someone assigned to a container vessel, tanker, or barge operations, material exists out there to get them acquainted with what to expect, but none exists for drillships; the same goes for drilling operations on land rigs—multiple materials exist, but no material for drillships; this is the intent of this book. Details will not be covered on the functionality of specific equipment, but rather on how this equipment works together to make these units capable to operate the way they do. The book is split into five sections:

Marine equipment: In this section we will cover various marine-related equipment on board. The experienced mariner might find this section dry, since it is meant for people not versed with vessels. A cadet, engineer, or someone with extensive drilling experience will find this section useful.

Drilling equipment: Here we will cover the "mud cycle," alongside various pieces of drilling equipment very similar to what are on a land rig, except they are bigger and more powerful.

Subsea equipment: In this section we will cover various subsea-related equipment. A lot of this equipment works hand in hand with drilling equipment, but as the reader will find, a lot of this equipment is interwoven. Anyone new to drillships will find this useful.

Supporting operations: In this section we will cover other supporting operations that you will find on a modern drillship but might not necessarily be owned or operated by the vessel owner or regular rig crew. Without these, drilling operations could probably not run as smoothly or efficiently. In addition to these will be covered non-work-related items you would expect to find on modern vessels.

Technological advances / new developments: This last section covers the wave of the future and what advances are being made in the industry to make these units more efficient not only to increase productivity, but to make these units safer. The reader might find some areas a stretch, but forty years ago, cell phones were tales we read about in comic books.

The prototype vessel used throughout this book is a DSME (Daewoo Shipyard and Marine Engineering) 12000-class vessel. DSME is among the biggest builders of drillships worldwide, and most shipyards name the class of vessels after the shipyard; thus this class of vessel is referred to as the DSME 12000. Shipyards generally have a baseline of what a specific class of vessel will look like, and clients make certain specific changes on certain equipment. The general vessel layout will be the same. Following is the general vessel layout, with the descriptions and locations of certain compartments and where certain equipment will be located. The reader may have to refer to this general layout throughout the book.

Vessel Overview

Stern, which will contain the following:
engine rooms, engine control room, aft thrusters (all below deck), and the casing funnel

Riser deck, which will contain the following:
riser and deck storage, ROV, well test equipment, fuel tanks (below deck)

Derrick module, which will contain the following:
rig floor and doghouse, subsea office and workshops, BOP's moon pool, shale shaker rooms, and process rooms

Mud module, which will contain the following:
mud pump rooms, pipe deck, warehouse/storerooms, active and reserve pits, workshops, dry bulk tanks room (below deck)

Bow/accommodation, which will contain the following:
offices, navigation bridge, rooms, gymnasium, forward machinery space, forward thruster rooms (below deck)

INTRODUCTION

History of Drillships

Offshore drilling has been around since the 1930s, when swamp barges were used to drill worldwide, and it was not until 1956 that the first drillship, named the *CUSS I*, was developed by a consortium of Continental, Unocal, Shell, and Superior Oil. The *CUSS I* had four thrusters at each corner of the vessel and measured 260 × 48 ft., with a gross tonnage of 3,000 tons, designed to stay within a 590 ft. (180 m) radius. Rather than the use of a GPS system (the same system used on cell phones today) for station keeping, the *CUSS I* used a combination of sonar measurements from preinstalled buoys, or the operator could adjust the vessel position by sight from these buoys.[1] After the *CUSS I* came the *Eureka*, which also used some upgraded control systems for better station keeping, but both the *Eureka* and *Cuss I* predominantly collected core samples from the seabed in water depths up to 350 ft. when other vessels were drilling in 200 ft. of water depth.

The first drillship to use a combination of riser and blowout preventer (BOP) was the *Sedco 445* (later renamed the *Deepwater Navigator*), which came into operation for Shell in 1971 and was capable of drilling in water depths of 7,000 ft. in the Gulf of Mexico. After the *Deepwater Navigator* came modern drillships using GPS satellites and beacons on the seafloor for station keeping, pioneered byTransocean, with the *Discoverer Enterprise* (Enterprise class) from 1999 to the mid-2000s. These vessels have dual-activity capability (two top drives on the rig floor) and can drill in 10,000 ft. of water with a BOP at 10,000 psi capacity. The final leap came in the early 2010s, with vessels capable of drilling in 12,000 ft. of water, with heavier hook loads for the top drives at more than 2,500,000 lbs. and 15,000 psi rated BOPs.

[1] Dynamic-positioning.com/history-of-dp

As technology evolved, the focus became redundancy of operating systems to reduce any downtime and backup capacity for batteries, engines, and pumps to manage drilling operation running 24/7. These modern rigs measure more than 750 × 130 ft., with a gross tonnage of 68,000 tons, and typically stay well within 1 ft. of the well, even in adverse weather conditions (45 kts. of wind and 10–12 ft. seas and swells)—a stark difference and improvement from the *CUSS I*.

Marketing and Contracts

Drillships are owned and operated by offshore-drilling contractors (e.g., Transocean, Seadrill, Valaris, COSCO, and Noble). The crew on board the drillships are typically employees of the offshore-drilling contractor. The unit gets contracted to drill a well (or series of wells) on the basis of the drilling program provided by a client. Contracts can be executed either for a particular number of wells or a certain period of time (time charter). This is typically what the market had seen until the effects of COVID-19 in 2020 snowballed into a slump in oil pricing and the eventual bankruptcy of multiple drilling contractors. A lot of these contractors emerged from bankruptcy, whereby assets (the drillships) are managed by different contractors, or vessels are leased on bareboat charters, with the manning and maintenance being handled by a different entity. This is highly unusual in the industry, given the multiple moving parts and cost of running the equipment on these high-specification rigs. Clients for drilling ships typically fall into three categories:

- national oil companies (NOC): These are companies that are owned and operated by a state/government
- major nationals / international oil companies (IOCs): These are well-established oil and gas companies that have a global footprint.
- independents: These are relatively small oil and gas companies who focus in a particular region of the world or sector of the offshore industry.

Depending on market conditions, the contract for a drillship can vary from one well for a few months and for as long as ten years, though ten-year contracts are extremely rare; two-to-three-year contracts with options for renewal typically reflect a stable market. When oil prices are high, drilling contractors get their rigs at favorable rates and terms to them; in the late 2000s, with oil prices over $100/barrel (bbl.), some drillships were contracted for as high as $750,000/day. When oil prices are low, the rate drops significantly; after the 2014 oil crash, some drillships got contracted for as low as $125,000/day. Utilization in a particular region also plays a key role on day rates. Higher utilization (percentage of available rigs in a region currently in use) implies there are fewer rigs available, and rates tend to increase; the same is true for lower utilization rates. Typically, when utilization rates have been north of 80%–85%, rates tend to be favorable for drilling contractors. On the basis of this cycle, at low oil prices, independents and NOCs tend to have more drilling programs and tend to cover most of the market, and when prices rise, the IOCs make up most of the market; as such, finding a balance among NOCs, independents, and IOCs is what most contractors strive for. The cycle the offshore-drilling industry follows is almost in line with that of oil prices (or slightly lags behind). As oil prices rise, this is typically reflected as higher profits for companies, and, in turn, more funding goes toward exploration programs for those companies.

The frontiers of deepwater drilling have typically been referred to as the Golden Triangle, comprising West Africa, the Gulf of Mexico, and Brazil. However, other areas that have seen an uptick in exploration activity with success include East Africa, the eastern Mediterranean region, India, Australia, Guyana, and the southern Caribbean.

Specifications for any type of drillships can typically be found on any of the drilling contractors' websites, such as the ones listed (see p.12).

Regulatory Compliance

Drillships fall into two categories when it comes to regulatory compliance: what regular seagoing vessels have to comply with, and what drilling rigs have to comply with.

As a vessel, there are two governing bodies; namely, the flag state and the classification society. In addition to these two, the port state control might have additional requirements for the vessel to meet, even though most, if not all, of what the port state control requires will already have been met by complying with the flag state or classification society (or both). The port state control will depend on the operating region of the vessel.

DEEPWATER INVICTUS

ULTRA-DEEPWATER DUAL-ACTIVITY DRILLSHIP

For additional information please contact:
Transocean
Marketing Department
1414 Enclave Parkway
Houston, TX 77077, USA
Phone: + 1-713-232-7500
Fax: + 1-713-232-7880
marketing@deepwater.com
www.deepwater.com

General Description

Design / Generation	DSME 12000 Ultra Deepwater Drillship
Constructing Shipyard	DSME (South Korea)
Year Entered Service / Significant Upgrades	2014
Classification	DNV X1A1
Flag	Marshall Islands
Dimensions	781 ft (238 m) long x 138 ft (42 m) wide x 62 ft (19 m) depth
Drafts	39 ft (12 m) operating / 29 ft (9 m) transit
Accommodation	200 persons
Displacement	114,640 st (104,000 mt) operating
Variable Deck	22,636 st (20,535 mt) operating / 22,636 st (20,535 mt) transit
Transit Speed	up to 12.5 knots
Maximum Water Depth	12,000 ft (3,658 m) designed / 10,000 ft (3,048 m) outfitted
Maximum Drilling Depth	40,000 ft (12,192 m)

Drilling Equipment

Derrick	Dual Aker MH Pyramid Dynamic Derrick 213 ft (65 m) x 69 ft. (21 m) x 52 ft (16 m)
Hookload Capacity	(Main) 2,800,000 lbs. gross nominal capacity (Aux) 2,200,000 lbs. gross nominal capacity
Drawworks	(Main) AKMH Wirth GH 9000 EG-AC-1G, 4 x AC motors; 9,000 hp, 1,553 mt with 16 x 2-1/4 inch lines strung lines (Aux) AKMH Wirth GH 6000 EG-AC-1G, three AC motors; 6,000 hp, 1,256 st (1,140 mt) with 14 x 2 inch lines strung
Compensator	(Main) Aker MH Crown Mounted Compensator with Active heave compensation, 750 st (680 mt) with 25 ft max stroke. (Aux) Aker MH Crown Mounted Compensator with Active Heave, 500 st (454 mt) with 25 ft max stroke.
Rotary Table	(Main) AKMH Wirth RTSS 75-1/2 inch hyd. 1250 st (1,134 mt) (Aux) AKMH Wirth RTSS 60-1/2 inch hyd. 1,000 st (908 mt).
Top Drive	(Main & Aux) 2 x Aker MH MDDM-1250-AC-2M 1,250 st (1,134 mt) tripping load, 945 st (857 mt) drilling load, 101,000 ft.-lbs. max continuous torque at 113 RPM, 0 - 280 max RPM, 2 x 1,050 hp ABB AC motors.
Tubular Handling	2 x Aker MH Hydraulic Iron Roughneck MH 1899, multi frame, tubular range 3-1/2 inch to 9-3/4 inch. Aker MH racking systems tubular range 3-1/2 inch to 14 inch. 2 x Vertical Pipe Racking Systems including 15 st (14 mt) bridge cranes with lower guiding arms. 1 x Horizontal Tubular Feeding Machine to handle single joints drill pipe and casing or casing stands up to 90 ft.
Riser Feed	1 x Horizontal Riser Feeding Machine for 21 inch x 75 ft (23 m) Riser Joints.
Mud Pumps	4 x AMH/Wirth TPK-7-1/2 x 14, 2,200 hp 7,500 psi
HP Mud System	Rated for 7,500 psi.
Solids Control	8-x Derrick Dual Pool (DP626); High G shale shakers

Revision Date: 30 September 2021

Power & Machinery

Main Power	6 x HHI HiMSEN H32/40V V-type diesel engines rated 7,000 kW, 720 rpm, each driving 1 x 8,125 kVA AC generator
Emergency Power	One Caterpillar 3516B V-type diesel engine rated 1,780 kW, 1,800 rpm driving 1 x AC generator
Power Distribution	3 x Siemens NXPlus C Plus, 11 kV Switchboards with AKA Advanced Generator Protection

Storage Capacities

Fuel Oil	51,280 bbl (8,153 m3)
Liquid Mud	9,745 bbl (1,549 m3) active / 10,562 bbl (1,679 m3) reserve
Base Oil	5,031 bbl (800 m3)
Brine	5,031 bbl (800 m3)
Drill Water	18,869 bbl (3,000 m3)
Potable Water	9,359 bbl (1,488 m3)
Bulk Material	(mud + cement) 28,251 cu.ft. (800 m3)
Sack Storage	10,000 sacks

BOP & Subsea Equipment

BOP Rams	(Primary) Cameron 18-3/4 inch, 15,000 psi 7-ram preventer; (2 x TL Doubles + 1 x TL Triple).
BOP Annulars	(Primary) 1 x Cameron Dual DL 18-3/4 inch (2 x elements), 10,000 psi annular preventer
2nd BOP Stack	(Secondary) Cameron 18-3/4 inch, 15,000 psi 7-ram preventer; (2 x TL Doubles + 1 x TL Triple) with 1 x Dual annular
BOP Handling	BOP crane 2 x 275 st (249 mt) main hoists and 2 x 16.5 st (15 mt) service hoists; with BOP Trolley rated 595 st. (540 mt) 2 x 551 st (500 mt) BOP storage stands with sea-fastening.
Marine Riser	Cameron Load King 21 inch, 3,500 kips (H Class), 75 ft (23 m) long per joint
Tensioners	8 x Aker MH Dual wireline riser tensioners, 250 kips each. Total capacity 4,000 kips with 12.5 ft (4 m) stroke 50 ft (15 m) max stroke.
Diverter	21-1/4 inch 500 PSI Type CSO diverter with 16 inch flow line
Tree Handling	1 x 165 st(150 mt) Xmas tree trolley with 2 x 165 st (150 mt) Xmas tree skid carts supplied with room to store up to 4 x carts. Xmas trees can also be lifted or handled with the 181 st. (164 mt) Active Heave Compensating Subsea Crane.
Moonpool	82 ft. (25 m) x 33.5 ft. (10 m) Outfitted with a two piece Moonpool Guide Array for improved transit characteristics.

Station Keeping / Propulsion System

Thrusters	6 x (three at forward, three at aft) Rolls Royce 5,500 kW variable speed, fixed pitch, fully azimuthing, underwater demountable thrusters.
DP System	Kongsberg DPS-3 rated for water depths up to 12,000 ft. (3,658 m)

Cranes

Cranes	3 x 110 st (100 mt) NOV knuckle-boom cranes model OC4000KCE with semi-automated riser and tubular handling attachments.
AHC Subsea Crane	1 x 181 st (164 mt) NOV Active Heave Compensation knuckle-boom boom crane for handling subsea equipment up to 12,000 ft (3,658 m) WD.

Other Information

Helideck	Rated for Sikorsky S-61 & S-92 helicopters
Additional Features	MPD Ready

Vessel Specification

DEEPWATER INVICTUS

ULTRA-DEEPWATER DUAL-ACTIVITY DRILLSHIP

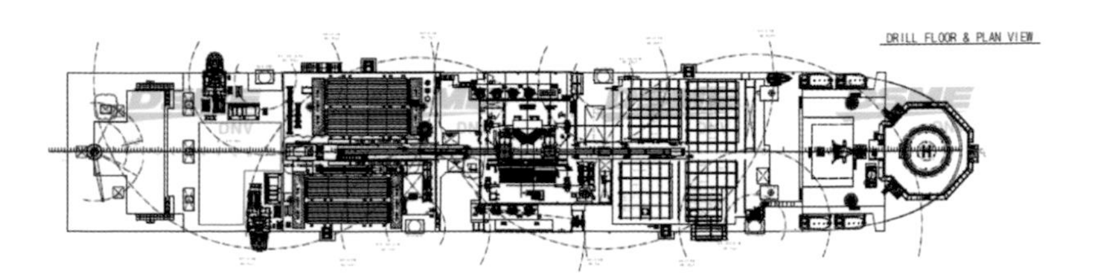

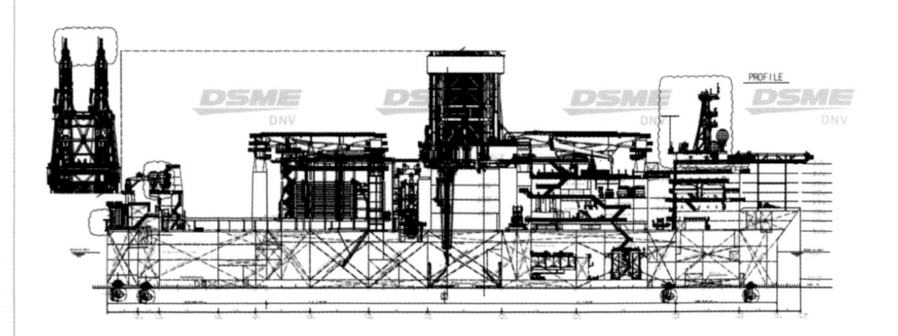

BOUNDLESS

SERVICE FOCUSED
DATA DRIVEN
PERFORMANCE ORIENTED

Revision Date: 30 September 2021

These specifications are intended for general reference purposes only, as actual equipment and specifications may vary based upon subsequent changes, the contract situation and customer needs. All equipment shall be operated and maintained at all times, in compliance with Transocean standard operating manuals, policies and procedures, and within its stated operational limits or continuous rated capacity, in order to assure maximum operational efficiency.

Patent Notice: This rig, its systems, components, and/or equipment in use on this rig, may be protected by one or more US and/or foreign patents.

The flag state is also known as the registry of the vessel; this is the country where the vessel is registered and hence will fly the flag of that country. On the high seas, vessels are required to comply with standards that are agreed upon globally through the IMO (International Maritime Organization), and it is the duty of the flag state to reinforce those rules for the vessels under her registry. Some requirements relate to the number of officers that must remain on board as a minimum and the amount of safety equipment that must be on board, such as the number of lifeboats and life rings, the frequency of fire drills, and even the proper operation of certain equipment, such as the ship's radios. These are just a handful of criteria that must be met by the vessel owner. Flag state inspections are carried out at least once a year and usually last five to seven days. If there are any infractions of rules and regulations while the vessel is operating, it is the duty of the captain to notify the flag state of what has occurred (for example, failure of a critical piece of safety equipment), and the flag state will either force the owner to stop operations until it is rectified, or provide the owner with a temporary issuance to operate for a certain period of time.

The classification society deals mostly with the technical aspect of the ship, and its main focus is easily remembered as "hull and machinery." Anything dealing with hull and machinery of the vessel will be covered by the classification society, such as cranes, condition of tanks, operating parameters of the dynamic-position system, condition of the top drive and drilling equipment, and all equipment that falls under hull and machinery. Through the International Association of Classification Societies (IACS), a lot of technical rules are agreed upon between all classification societies to ensure homogeneity. Members of the IACS are represented worldwide and are not country specific. Rules followed through the classification are also covered on drill rigs, since these societies also govern offshore structures. The classification society visits every year, and every five years an extensive survey is done, which can go for a few weeks to months, depending on what needs to be addressed by the ship owner.

The port state control represents the sovereign state/country where a vessel is operating. It is rare that port state control will have specific rules not covered by the class society or flag state (or both), but given that a vessel is operating within the territorial waters of a country, it is the responsibility of the port state control that certain specific rules are met. These specific rules are provided to the drilling contractor before the vessel enters its waters, to ensure no specific rules are broken. For example, certain countries have specific rules regarding the discharge of certain fluids overboard and will ensure that these specific rules are adhered to.

On the drilling aspect, a lot of requirements that are followed are set by the International Association of Drilling Contractors (IADC). Though the IADC is not a governing body, they set standards that are followed in the entire industry. Most drilling requirements are regulated depending on the operating region of the vessel. In the United States, the EPA (Environmental Protection Agency) has requirements that have to be met regarding what can be discharged overboard during drilling operations; this is reinforced through the National Pollution Discharge Elimination System (NPDES). In addition, the Bureau of Safety and Environmental Enforcement (BSEE) reinforces safety regulations as they pertain to offshore drilling, safety regulations that are maintained in the Code of Federal Regulations (CFR). Though different states will have different requirements for drilling operations, multiple items are very similar when it comes to the operation of the drilling unit. A lot of the differences come into play regarding what can or cannot be discharged overboard and what safety standards are permissible depending on the operating region.

Offshore Hierarchical Structure

The offshore structure is broken into two parts: one section where people report to the captain/OIM who represents the vessel owner/operator, and another section where people report to the company representative (the client who is leasing the drilling unit), otherwise known as the "company man." This reporting structure is for a particular drilling contractor, but you can expect to see the same reporting structure industry-wide with minor modifications. Though there are two different reporting lines for crew members, the Captain always has the ultimate responsibility of the vessel and this offshore hierarchical structure strictly relates to reporting for the ease of information flow.

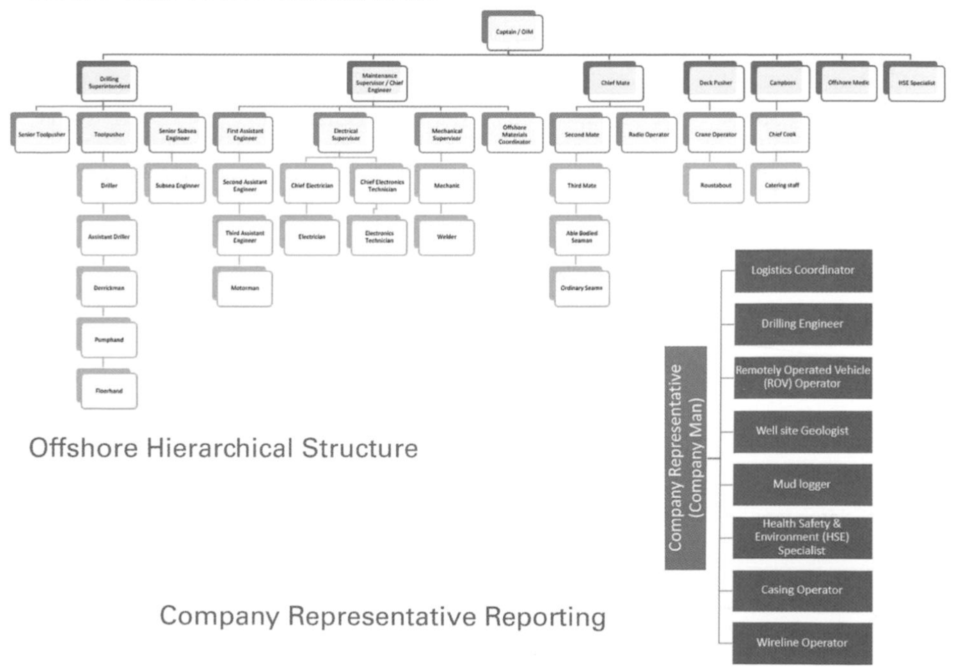

Offshore Hierarchical Structure

Company Representative Reporting

2 MARINE EQUIPMENT

2.1. Firefighting Systems

In the maritime environment, the vessel crews are the sole personnel responsible for fighting any fire on board a vessel. As such, certain crew members are assigned to fire teams on the basis of their qualifications and level of training. Here are the different systems used for fighting and containing a fire offshore.

2.1.1. Fire Main

The fire main is a piping system made up of salt water, taking its suction from the sea. The piping goes into every space aboard the vessel and eventually forms an entire loop, and it is sometimes referred to as the "ring main" or fire main. Think of this as forming a complete circle around the entire vessel.

There are typically three main fire pumps that support the entire system, with one pump capable of supplying enough pressure on the entire system while running. The fire main pumps take their suction either through an independent sea chest or through the main saltwater line in which the fire pump is located. There is usually one main fire pump in the bow and two on the stern of the vessel. The fire main system remains under constant pressure with the help of two jockey pumps and a hydrophore tank (jockey pumps are smaller compared to fire main pumps). The vessel operator sets certain parameters, at which point the jockey pumps automatically start to maintain a certain amount of pressure on the fire main by pressuring up the hydrophore tank. The hydrophore tank contains a mixture of pressurized air and salt water, thereby keeping water in the fire main under a certain amount of pressure. As such, the fire main is a

Jockey pump with hydrophore

pressurized system, with salt water always ready at any location of the vessel when needed to fight any fire.

2.1.2. Sprinkler System

The sprinkler system is also a pressurized system, but unlike the fire main, it is made up of fresh water and covers only the accommodation/main house of the vessel. Similar to frangible bulbs you find in hotel rooms or commercial buildings, once a certain temperature is reached, the frangible bulb is broken, letting pressured water into the space. A set amount of fresh water is held in a tank, and most sprinkler systems will have a saltwater pump that can provide salt water into the system if more water is needed. This backup saltwater sprinkler pump is usually in the forward machinery space.

2.1.3. Foam System

Foam systems cover as many areas as the vessel owner might want, but for practical purposes the entire vessel cannot have an active foam system. Each foam system is composed of AFFF (aqueous film-forming foam) held in storage tank(s); there is typically one tank forward of the vessel and one aft. Through independent foam pumps, this AFFF is mixed with salt water from the fire main system to form foam. Vessels are known to have an independent foam pipeline to cover certain areas. Drillships that have the capability of carrying crude oil will have a foam system that covers the decks above the crude oil tanks and loading and unloading stations, and alongside, the ability to reach the associated piping through the discharge manifolds, generally located on the stern of the vessel.

Helidecks will also be covered by either the same foam system or a dedicated foam tank and pump(s) adjacent to the helideck for faster delivery of foam. The same system covering the helideck will generally cover the associated helicopter fuel tanks and refueling station.

Helideck fuel tank with deluge system

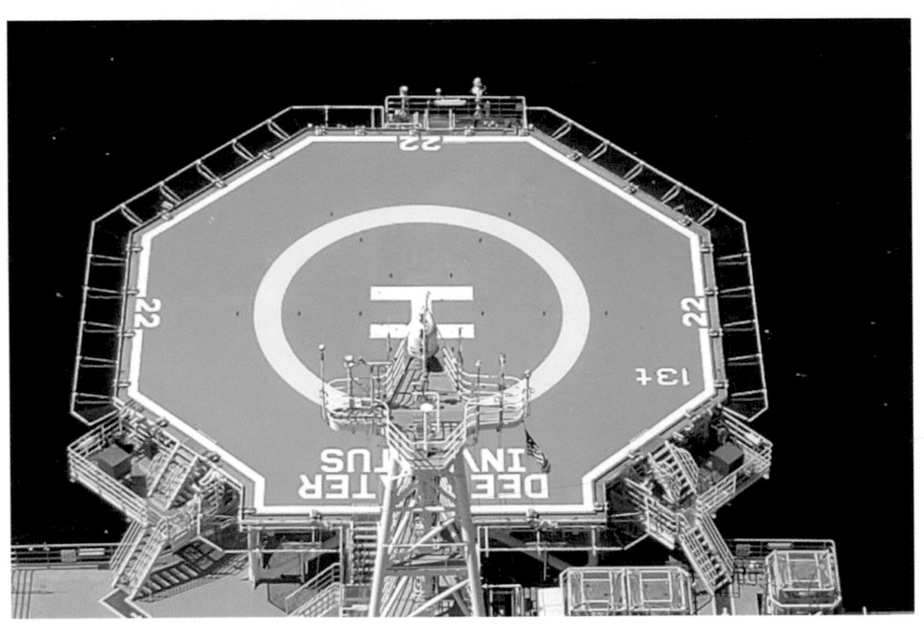

Helideck with monitors

The same foam system delivering foam to the tanks will have monitors on three corners of the helideck capable of delivering foam. Some helidecks are designed to have sprinklers that pop off from the helideck and deliver foam instead of having monitors around the vessel.

Helideck fuel tank with deluge system

Other areas covered by foam include engine rooms and their associated purifier rooms. Activation of the foam system can generally be done remotely from a foam pump room, the bridge, or a locker where the fire teams will muster in case of a fire.

Inergen piping, activation point and wiring

Inergen bottles

For local use of foam in firefighting, portable foam tanks (5 gallons each) are used by fire teams with the aid of special connections and the use of the fire main system; a suitable foam mixture can be obtained and used locally during firefighting.

2.1.4. Fixed CO_2 / Inergen System:

Fixed CO_2 or Inergen systems are used in most machinery spaces and, as described, are fixed systems. The industry is gradually phasing out CO_2 systems, since the material is lethal and Inergen is not. Inergen is composed of nitrogen alongside an inert gas such as helium or argon, with an oxygen content low enough that does not support combustion, but not too low to be lethal. Inergen systems have an independent piping from their main bottle banks to their fixed discharge nozzles. No two vessels will have both systems (CO_2 and Inergen), so the description can be used to cover either system. Where a difference does arise, it will be pointed out.

The Inergen system has an independent compartment containing all bottles used to cover the dedicated areas on board. From this same locker, the crew members are capable of discharging Inergen in a particular space covered by the system. In addition to this activation point, there exist local activation points outside the area covered. As per regulatory requirements, once the crew member discharges the Inergen under the captain's orders only, a siren and visual indication will notify any member of the imminent discharge of the system. The following list covers areas that typically contain fixed firefighting systems, but is by no means an exhaustive list: thruster rooms, engine rooms with associated casing rooms, purifier rooms, switchboard rooms, electrical rooms, paint lockers, engine control rooms, certain workshops, warehouses, and cement rooms, to name a few. Vessels typically carry enough Inergen or CO_2 to cover only the largest space in the vessel; if discharged at any point, these bottles will either need to be replaced or filled back up.

2.1.5. Deluge System

Deluge systems are branched off the fire main system but are also fixed firefighting systems. The deluge system provides a curtain of salt water via fixed nozzles around the area it covers. These systems normally are found around the rig floor, the moon pool, and well test areas / deck. During flaring operations, a deluge-like system is also provided to absorb the heat emitted during flaring. Either through a local or remote activation point (or both), a valve is opened, allowing salt water from the fire main system to form a curtain of water. Once there is a drop in pressure from the fire main, the main fire pumps can be started to keep supplying water to the deluge system.

Left: Deluge system activated

Middle: Moonpool deluge piping

Bottom left: Moonpool deluge piping

Bottom right: Deluge nozzle

2.1.6. Water Mist System

Water mist is another fixed firefighting system on board, which consists of a fine freshwater mist. The water mist system is run by a fresh water pump(s), with suction taken from the potable water tanks. Areas covered by water mist typically include the engine rooms, purifier rooms, and thruster rooms (when separate from the engine room). Activation can usually be done locally outside the space or remotely from the bridge via panel. Due to the fine water particles, electrical equipment does not get damaged, and this is usually used as the first line of defense in firefighting on board. Even with the water mist system activated, machinery and equipment in the space can be kept running without any damage.

Watermist Pump

Watermist piping over engine

2.1.7. Fire Extinguishers

Fire extinguishers will be in conspicuous areas around the entire vessel, and most likely in every space of the vessel. The type of fire extinguisher in an area will be determined by the type of fire most likely to be encountered in that compartment, or an adjacent compartment. A variety of sizes and types of extinguishers are used on board, including water, dry chemical powder, wet chemical, carbon dioxide, and foam.

2.2. Marine Diesel Generators and Power Distribution

Diesel generators are the only source of power for the entire vessel, and as such, without any power, no equipment can operate. For ease of power distribution and control, the generators on board are diesel electric, meaning that a generator provides electrical power through an alternator, which, through transformers and electrical wires, supplies power to all units on board the vessel. A lot of auxiliary equipment assists in running these diesel generators, such as fuel pumps, lube oil pumps, saltwater pumps, and freshwater pumps.

Modern deepwater drillships have three separate engine rooms, typically consisting of two diesel generators each usually rated between 5,000 and 7,500 kW generating 11,000 volts of 60 Hz power. These three engine rooms provide power to separated buses, and through bus tie-breakers these three separate buses are connected to form one unified bus. Classification societies require this separation for ease of isolating faults and preventing one side being affected by the other; the primary purpose of a deepwater drillship is to stay in one position while a well is being drilled, so redundancy is critical (redundancy in these systems is discussed later). Step-down transformers reduce the 11,000 volts all the way to 690 volts to power pumps and heavier equipment, and eventually to 110 volts to supply small equipment, such as a desktop, and further down to 24 volts for even-smaller equipment, such as solenoids.

This power is distributed by cables that run around the vessel through areas designated by the classification societies. The stern of the vessel contains switchboards and transformer rooms through which equipment supplied in the stern will be fed. Similarly, the drilling package (area surrounding the drill floor and drilling equipment) also has an electrical room, which will have switchboards and transformers supplying power to mud pumps, top drives, and all power related to the drilling package. The bow will also contain a transformer room and switchboard room, through which equipment in the forward part of the vessel is supplied. This compartmentalization makes it easy to secure and distribute power to various equipment around the drillship, though certain equipment aft might be fed from the forward switchboard either as a backup or redundancy (the fact that equipment is forward does not necessarily mean its power distribution is coming from a forward switchboard). The electricians and marine engineers will have an entire list of what set of equipment is fed by a particular switchboard for every single space. This is critical for troubleshooting and isolating equipment when maintenance is being carried out.

In addition to the main diesel generators, vessels will be equipped with an emergency diesel generator (EDG) with a much-lower capacity (usually between 1,500 and 3,000 kW). Similar to the main diesel generator, the EDG provides power to a bus called the emergency bus. As per flag state and class regulations, certain critical equipment will be fed from this emergency bus, some of which may include a bilge pump, fire pump, foam pump, ballast pump, and certain lighting around the vessel. During normal operations the emergency diesel generator is not used, and as such the emergency bus is powered from the main bus through step-down transformers and circuit breakers. EDGs are designed to automatically start when the EDG realizes (through a programmable logic controller) that there is no power to the emergency bus.

Engine room with overview with MDGs

Switchboard room

2.3. Fuel Oil and Lube Oil

2.3.1. Fuel Oil

The fuel oil system is an auxiliary system used primarily to supply fuel to the main and emergency diesel generators. Piping around the vessel allows for the transfer of fuel to lifeboats, fast rescue boats, and equipment run by other service providers that support the drilling operation. This description focuses on how the fuel oil system is used for supplying fuel to the diesel generators. Fuel transfer to other units is simply completed by a transfer pump from the engine room through fuel pipes to a nozzle on deck.

The fuel system is compartmentalized so that each engine room (port, starboard, and center) can receive fuel from tanks from that particular engine room. Crossover valves exist for ease of transferring fuel from one engine room to another. One engine room consists of a main fuel-holding tank, a day tank, and a service tank alongside fuel oil purifiers. Fuel received from ashore (through a supply boat during normal operations) is taken into the main storage tanks, usually capable of holding about 2,500 m^3 of fuel. From this main day tank, fuel is transferred to the settling tanks (capacity of about 200 m^3), which are small enough for engineers to accurately measure how much fuel gets used per day. From the day tank, the fuel is passed through the purifiers and kept in the service tanks. As the name indicates, purifiers are used for getting rid of any impurities in the fuel.

Once the engine in one engine room is given the start command by the engineer, fuel is pumped from the service tank to either one or both engines in the engine room. Each engine room consists of a main fuel pump and standby pump, allowing engineers to be able to isolate one pump for maintenance without having an effect on the functioning of the engine in that engine room.These two fuel pumps are electrically driven, and in addition, there exists a pneumatically driven fuel pump.This pneumatically driven fuel pump is run by air stored in a reservoir tank (otherwise known as an accumulator tank) and is used only during blackout situations (when power is completely lost on the vessel). Given that there is no power on the entire vessel during a blackout and the electrical pump cannot run, the pneumatic driven pump provides a means of supplying fuel to the engine in the machinery space.The relatively small capacity of the pneumatically driven pump will typically allow for only one engine to run for a period of time, after which other equipment can be brought back online through electrical power from the main generator online.The capacity of the pneumatic pump will vary from one vessel to another.

Fuel oil purifier

This depicts only one engine room; the same can be replicated for all three engine rooms. For emergency purposes, there exist quick-closing valves, again pneumatically operated by air stored in a reservoir tank. These quick-closing valves (QCV) shut down the flow of fuel from service tanks and day tanks and eventually starve the engine of fuel, thereby shutting the engine down. These are used in emergency situations such as fires, when fuel supply needs to be shut off at all cost.

Reservoir for quick-closing valves

Quick closing valves

2.3.2. Lube Oil

Lube oil is used for the lubrication of the main diesel generators (MDG). Drillships can have independent main storage tanks (aft) or one main storage tank that feeds all systems. The former system is most commonly used; however, the main risk involved is getting contaminated lube oil in the main tank and eventually the entire system. Notwithstanding, from the main tank lube oil is fed into the MDG via lube oil pumps after going through a lube oil purifier; each MDG runs with its own lube oil purifier. From the MDG a sump captures lube oil, after which it is recirculated or sent to a different storage tank, from where it is discarded.

Lube oil purifier

2.4. Thrusters

Thrusters are the only propulsion system used on board modern drillships, with three forward and three aft, each in a triangular pattern. Thrusters can either be a retractable type (able to be lowered into the water or raised in their respective machinery space) or nonretractable, in which case divers and cranes will be needed for pulling out the thruster. They are fixed-blade-azimuth (can turn 360°) thrusters rated for about 7,000 hp, drawing about 5,500 kW at full capacity and producing as much as 100 tons of force each. They measure about 5.5 m (18 ft.) below the baseline (bottom of the vessel).

The two main cooling systems for thrusters include salt water and fresh water. Fresh water is used for cooling air-conditioning units in the space, lube oil systems, hydraulic oil systems (steering power packs), transformers, and VFDs (variable-frequency drives). The salt water is solely used in the medium as a heat exchanger.

The forward thrusters are in independent compartments, with each of their auxiliary units contained in the same compartment. The saltwater sea chest and pumps will be in the same space alongside the freshwater pumps and the tank supplying the fresh water. The only items that might be common to each of the thrusters forward could be the main tank supplying lube oil and hydraulic oil; note that this is just for a main holding tank, since each space will have its own expansion tank for these systems (both lube oil and hydraulic oil). One main tank can be used to supply each of the systems, but each thruster will have its own lube oil and hydraulic oil pumps.

The aft group of thrusters is aft of each engine room and is either separated by a watertight bulkhead or just by a fire door; the extent of separation depends on the construction of the vessel and the owner's preference for redundancy. Each engine room will typically share a common saltwater system with the associated thruster. For example, the saltwater system on the port side of the engine room will also circulate saltwater cooling for the thruster on the port aft corner of the vessel; the same concept is used for the starboard and center engine rooms. Similar to the saltwater system, freshwater systems will be common for the engine room associated with a particular thruster.

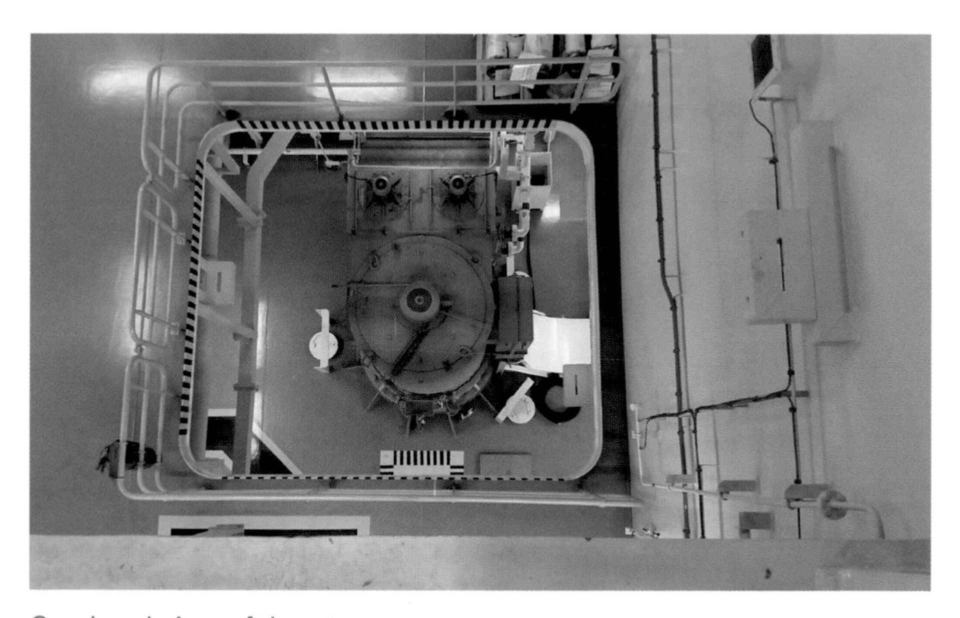

Overhead view of thruster

Thruster view

Thruster on deck

2.5. Air Compressors

The discussion of air compressors covers two main sections: the main rig air compressors and the start air compressors for the MDGs. Rig air compressors supply pressurized air around the vessel and are used for various equipment, while start air is solely for starting the MDGs.

2.5.1. Rig Air Compressors

Rig air compressors are generally all in the forward machinery spaces and can vary in number, ranging from three to as many as five; it all depends on how much equipment on board uses air for functionality. Some vessels will have rig air compressors situated in the engine rooms or in separate rig air compressor rooms. Rig air compressors are used to supply all equipment aboard (excluding the main diesel generators), such as dampers (see "HVAC: Heating, Ventilation, and Air-Conditioning") and valves that are pneumatically actuated, and the transfer of bulk material (cement, barite, and bentonite). The distribution of rig air is aided by the use of air reservoirs (also known as accumulators) around the entire vessel and typically close to the equipment being supplied. For example, dampers that control the center engine room will have their reservoir tank adjacent to the center engine room, or in some cases within the center engine room. Once the air reservoir gets below a preset limit, the air compressor automatically starts to fill the tank and stops at a preset pressure. In addition to this, a common rig air line will go around the entire vessel, to which crew members can insert fittings for general deck work, such as needle gunning, or whatever small tool might need rig air; rig air typically operates at about 8 bars (about 115 psi).

Four rig air compressors

2.5.2. Start Air Compressors

Each engine room will contain an independent air compressor solely for starting the engines in that engine room. As such, three separate engine rooms will have three separate air compressors with separate air reservoir tanks. These systems are generally isolated from each other, though valves and piping allow for air to be supplied from one engine room to another. Similarly, the rig air compressor and start air compressor systems are completely independent and stay that way during normal operations, but valves also exist for one system to supply another during emergencies (e.g., a crossover valve can be used so that the rig air supplies the start air for the MDGs).

Start air compressors

2.6. Ballast

Ballast water tanks are on each side of the entire vessel. Salt water from the sea is pumped into and out of them purposefully to keep the vessel at a particular draft mark set by crew members. The vessel has to remain at a particular draft, not only to keep the vessel stable but also because the drilling program for a particular well is based on a particular draft of that vessel. The modern drillship usually has anywhere from twenty-three to thirty ballast tanks with a total capacity of about 60,000 tons of salt water.

The main sea suction for the ballast tanks will be either in a dedicated cargo pump room or the bottom level of the center engine room. In these rooms there are two sea chests supplying two ballast pumps that can be crossed over. From this space, the piping system allows seawater to be pumped into any ballast tank around the entire vessel, or overboard when pumping water out of tanks. In addition to the ballast pumps, ballast water treatment systems are used when pumping ballast water into or out of the tanks, depending on the vessel's location. The ballast water treatment system essentially "cleans" the water, preventing these ballast tanks from being used as a means of transporting certain organisms from one region of the world to another. Ballast tanks can also be used as a means of pumping out water from the machinery space in case of an emergency.

Ballast pump

BWTS from center machinery space

Ballast tank

2.7. Bilge

Bilge systems are separated into the forward machinery space and the aft machinery space. The bilge is the bottommost part of the vessel, where water from air-conditioning units can be drained into or waste from working on equipment will flow into, either intentionally or not (drainage from drill floor operations is covered in a different section, "Drain Holding Tanks"). Bilge wells (small containments within the bilge) on modern drillships will have floats that automatically set off an alarm (either in the bridge, engine control room, or both) and prompt an air-driven bilge pump to move the residue out of the bilge well and into a holding tank.

The control room allows the bilge system to be run either automatically (allowing valves and pumps to run automatically) or in "manual mode," allowing the operator to decide what to do once an alarm is received. From the holding tank, this residue is passed through an oily-water separator that monitors how much oil is in the residue, and if reached at or above 15 ppm (parts per million) the residue is not discharged into the sea. The residue containing high oil content gets sent into a different holding tank (typically one in the forward machinery space and one aft) and is then transferred into portable tanks that can be shipped to shore for safe discharge.

Each thruster room, purifier room, and lower sections of all machinery spaces will contain bilge wells alongside piping to pump the bilge into a holding tank. The holding capacity of these bilge holding tanks all depends on the vessel but usually does not exceed 70 m^3.

2.8. Saltwater and Freshwater Cooling

2.8.1. Saltwater Cooling

For ease of explanation of saltwater cooling systems, the discussion is split into saltwater cooling forward and saltwater cooling aft. It is important to note that all saltwater cooling systems provide a medium to cool off fresh water through heat exchangers. As such, salt water indirectly cools equipment through the use of heat exchangers. Without salt water, most equipment will continue running only as long as the freshwater temperature stays below a certain limit; this will depend on how "hard" an equipment is being run. If a thruster is run at 90% its rated capacity, its auxiliary equipment will get hot a lot faster than if it was run at 10%.

2.8.2.1. Saltwater Cooling Forward

In the forward part of the vessel, saltwater cooling is used for the forward machinery space and each individual thruster room. Each of the thruster rooms forward (thrusters 1, 2, and 3) contains its own independent sea chest from which two saltwater pumps from a common line draw suction, thereby making saltwater cooling for each thruster space completely independent. In the machinery space forward, there generally exist two sea chests (port and starboard), which have a main saltwater pipe connecting the sea chests. From this common saltwater pipe the fire main system will draw its salt water, alongside a saltwater pump cooling all topside equipment (rig floor equipment). Machinery operated in the forward machinery space is cooled by heat exchangers, such as rig air compressors, air-conditioning (AC) units for the main house (accommodation), smaller AC units for certain rooms, and a variety of auxiliary equipment.

View of forward machinery space

2.8.2.2. Saltwater Cooling Aft

In the aft part of the vessel (aft machinery space / engine room), vessels with three separate engine rooms will each have two independent sea chests, with the center machinery space sharing a sea chest with the port and starboard machinery spaces, making for a total of four sea chests for saltwater cooling (certain vessels have six sea chests–two in each engine room).

Saltwater cooling used for one engine room is shared with the thruster in the same engine room. Similar to the forward machinery space, two sea chests will have a common saltwater pipe connecting both sea chests. From here, saltwater pumps will take suction for use in heat exchangers used for cooling of engine auxiliaries (heat exchangers for fuel oil purifiers, lube oil purifiers, thrusters, VFDs, transformers, etc.). For engine rooms that contain a fire pump (typically port and starboard engine rooms), the fire main systems will also draw suction from the saltwater cooling pipe.

2.8.3. Topside Salt Water / General Salt Water

Topside aboard drillships generally refers to any equipment on the rig floor. A lot of equipment gets used during various drilling operations; as such, for washing down tools or cleaning certain spaces and tanks, general salt water is used. To make this available without using the same saltwater system for the fire main (solely used for firefighting), a separate and independent saltwater pipe runs to the rig floor and various areas where it might be used: shale shakers, mud pump rooms, etc. Aboard certain vessels this saltwater pipe runs through the entire vessel in conspicuous areas. Through regular valves and a different-colored pipe, salt water is at the disposal of anyone needing water for washing down or cleaning. This saltwater system usually takes its suction from the common saltwater pipe connecting the sea chest in the forward machinery space. However, this system can contain its own independent sea chest and suction pipe to make it fully independent if required.

2.8.4. Freshwater Cooling

Similar to the saltwater cooling system, the freshwater cooling system will be discussed in four categories:

- forward machinery space freshwater cooling
- aft machinery space freshwater cooling
- forward thruster freshwater cooling
- topside freshwater cooling

Before diving into each of these systems, it is important to understand that all freshwater cooling systems are closed-loop systems, meaning that if the piping is traced around from the discharge side of a pump all the way around the various cooling parts, you will end up on the suction side of the same pump, completing a full loop around the system. All systems have a reserve tank of fresh water typically called the expansion tank, where freshwater suction is taken from and discharged into. These tanks typically hold between 0.5 and 2.0 m^3 of water, depending on how big a cooling system is needed. Piping exists to fill up the tank if a leak occurs and more fresh water is needed, or if the system is completely drained out for maintenance purposes. Each freshwater cooling system will also have at least two heat exchangers where the "hot" fresh water is cooled, with salt water acting as the cooling medium.

2.8.4.1. Forward Machinery Space Freshwater Cooling

The forward freshwater (FW) cooling system is used to cool down any machinery in the forward machinery space, typically comprising

- rig air compressors
- air-conditioning units for the accommodation
- mooring winches
- any other air-conditioning units within the accommodation module

The system will comprise two pumps for redundancy, with only one running at any time. Any other equipment within the machinery space that will need freshwater cooling will be piped through this.

2.8.4.2. Aft Machinery Space Freshwater Cooling

Since the example drillship has three independent engine rooms, the discussion will be limited to one engine room, in this case a center engine room.

Fresh water for the center engine room will also be used to cool down the associated thruster, which in most cases will be thruster 6. Certain modern drillships further segregate the freshwater cooling system for an engine room with the thruster room, but this is a rarity, since it involves more piping, more pumps, and additional expansion tanks. Engine room freshwater cooling will cover part of the engine room, such as

- air-conditioning units for the associated switchboard room
- fuel oil cooler to the main engines
- lube oil coolers
- main engine start air compressor
- associated thruster cooling, such as the lube oil units, hydraulic units (used for steering), and transformer

Again, it is important to remember that this system is completely enclosed, with suction taken from the expansion tank. Similar to the forward machinery space, two pumps exist, with one running at any time and the other as standby.

2.8.4.3. Forward Thruster Freshwater Cooling

Since each thruster forward (1, 2, and 3) is completely independent, each has its own freshwater cooling system, consisting of their own expansion tanks, independent piping, and two freshwater pumps. They will typically be piped to cool

- thruster transformer
- air-conditioning unit
- lube oil cooler
- steering pump unit
- thruster motor

2.8.4.4. Topside Freshwater Cooling

Topside freshwater cooling is typically the biggest of all freshwater cooling systems aboard modern drillships, since it cools all the drilling equipment, and when dealing with dual-activity rigs (with two top drives), you get two of almost everything needing to be cooled. The system typically comprises at least three pumps, three heat exchangers, and a bigger expansion tank (usually 2.0 m^3). Pumps in the topside freshwater cooling are sometimes the biggest of all pumps aboard a drillship. In addition to drilling equipment being cooled by this system, any auxiliary equipment (such as cement units) that might need fresh water will be cooled by the topside freshwater cooling system. Certain equipment cooled by this system includes

- coolers for hydraulic units

- air compressors
- draw works oil cooler
- draw works AC motor cooler
- air-conditioning units for associated electrical rooms on the drilling floor
- air-conditioning units for workshops (mud control rooms, doghouse)

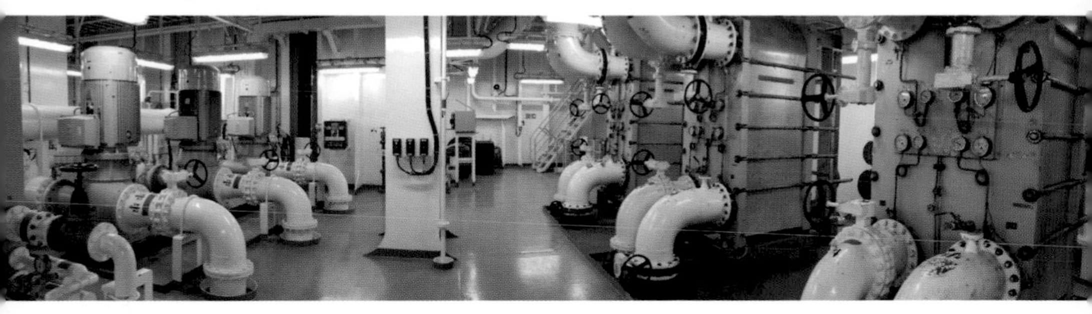

Fresh water cooling pumps

2.9. Drain Holding Tanks

To prevent any water or fluids from the deck going directly overboard, vessels today have been designed with drain holding tanks, mainly separated into nonhazardous and hazardous drain holding tanks.

Nonhazardous drain holding tanks will collect water that is spilled on deck through a drain system in areas that have been deemed nonhazardous, such as the bow of the vessel or the main house. These are essentially areas where no drilling operation takes place and there is very little chance of getting hazardous waste spilled on deck. From the drains around each deck and conspicuous areas around the vessel, water is collected in a "nonhazardous" tank and, after going through an oily-water separator, is discharged overboard.

Hazardous drain holding tanks will collect water or residue spilled on deck into a hazardous drain holding tank. Areas such as the rig floor, where a lot of mud gets spilled, and the subsea area will be collected in the hazardous drain tank. These tanks are generally inclined and contain baffles, so the mud settles at the bottom and the water eventually gets spilled over into separate compartments of the tank. Through this process the mud settles at the bottom and is eventually transferred into tanks and sent ashore for proper disposal, while the water gets pumped overboard after going through an oily-water separator.

For any residue that does not get past the oily-water separator, the water is sent to a sludge tank, usually the same tank containing residue from the bilge. From here it gets put in tanks and sent ashore for proper disposal.

Drain holding-tank room

2.10. HVAC (Heating, Ventilation, and Air-Conditioning)

To cover this topic in a thorough manner in regard to aboard drillships, it is necessary to describe areas that typically contain air-conditioning units or air-handling units and forced ventilation, which is achieved by dampers and fans.

Aboard any drillship, the biggest area using air-conditioning (AC) is the main house, which contains sleeping accommodation, offices, recreation rooms, a gymnasium, and a galley (to name just a few). The main AC unit supplying this area is typically in the forward machinery space just below the main house. Ventilation ducts from the machinery space throughout the accommodation help distribute cool air throughout all these areas. Individual spaces (such as a stateroom) will usually contain a thermostat with a heating unit, allowing personnel using that room to regulate the temperature for that room or space. The same goes for most spaces fed by this main AC unit. For the sake of redundancy there are typically two or three AC units, and depending on the manufacturer, one can usually suffice to cool the main house. Areas outside the accommodation that contain smaller air-handling units (AHU, used for cooling) are battery rooms, the warehouse, thruster rooms, the drill shack, electrical rooms, the engine control room, mud labs, switchboard rooms, and transformer rooms. This list is clearly not exhaustive, and any area that contains critical equipment needed to be kept at a certain temperature will contain an AHU outside or within the space, thereby keeping it cool. These areas are typically far away from the accommodation, and it would be impractical to have the same ducting supplying the entire vessel.

View of WCCU

General work areas that are enclosed (such as the pit room) are cooled by means of ventilation accomplished by having both supply and exhaust fans letting air circulate within the space. These fans can be shut off at any time, and the space needs to be either positively or negatively pressurized, so a combination of dampers are used to regulate the ingress or escape of air. To go any further in discussing ventilation aboard a drillship, it is important to cover the different classifications of hazardous areas as described by the classification societies:

> Hazardous areas are all those areas in which explosive gas or air mixture may normally be expected to be present in quantities which can require special precautions for the construction and use of electrical equipment and machinery. Hazardous areas are divided into zones depending upon the grade (frequency and duration) of release:
>
> a) Zone 0: in which an explosive gas atmosphere is continuously present or present for long periods. (Typical for continuous grade source present for more than 1000 hours a year or that occurs frequently for short periods).
>
> b) Zone 1: in which an explosive gas atmosphere is likely to occur in normal operation. (Typical for primary grade source present between 10 and 1000 hours a year).
>
> c) Zone 2: in which an explosive gas atmosphere is not likely to occur in normal operation, and if it does occur, is likely to do so infrequently and will exist for a short period only. (Typical for secondary grade source present for less than 10 hours per year and for short periods only).[1]

On the basis of this definition, certain areas are required to be either positively or negatively pressurized, depending on the hazardous area surrounding that particular space. For example, drill shacks (office containing the drilling operating system) are nonhazardous, while the area surrounding the drill shack (the rig floor) is classed as zone 2; for this reason the drill shack will be positively pressurized. In a similar fashion, a battery room in the accommodation is a zone 2, but the area surrounding that battery room (the accommodation) is a nonhazardous area, so that battery room will be negatively pressurized. Classification societies have many requirements for equipment in these hazardous areas, alongside requirements for air tightness, and for this reason, whenever there is a loss of pressure in this area, an alarm is triggered in the vessel management system / Integrated Automation System (VMS or IAS), notifying the operator that pressure has been lost and action needs to be taken.

It is very important for personnel to be aware of hazardous classifications and their implication to ventilation in spaces during drilling operations. Monitoring of all the spaces aboard a drillship is done from the IAS (see p. 43, 2.11). From the IAS, the operator can start and stop fans or shut off dampers. AHUs are typically not started from here, but when they shut down the operator also gets a notification. When there is a loss of pressure for a particular space, the operator is also notified. Given that ventilation is critical in firefighting aboard vessels, an automatic shutdown of ventilation and damper can be triggered when there has been a confirmed fire from the safety IAS without any intervention from the operator.

[1] DNV – GL Rules for Hazardous Classification

2.11. Safety and Safety IAS

This area discusses how fire and gas systems, emergency shutdown, and critical alarms are managed aboard drillships. The safety systems are split into three categories:

- safety IAS (Integrated Automation System)
- emergency shutdown (ESD)
- critical alarm and action panel

2.11.1. Safety IAS

Safety IAS (Integrated Automation System) provides a means of monitoring the fire and gas system onboard and comprises the operator stations in the bridge, the DP (dynamic position) backup room, and engine control room, all communicating via dual redundant networks to field stations that in turn communicate with detectors around the vessel. Uninterrupted power supply (UPS) provides backup power for sixty minutes to all this equipment, so in case of a blackout (total loss of power), there is no loss in the monitoring of the fire and gas system. Typically only the bridge operator station is "in command" of the system, with repeater monitors posted at other critical work areas such as the rig floor.

At a minimum, one can find the following types of detectors around the entire vessel: heat detectors, smoke detectors, flame detectors, hydrocarbon gas detectors, and hydrogen sulfide gas detectors. Certain vessels are equipped with nitrogen generators, and the area will contain an oxygen sensor. These detectors are in spaces depending on the expected exposure for that space, on the basis of requirements put forward by the classification society (an extract of what is required is listed).

These systems can be run in automatic or manual mode. If the system is operated in automatic mode, a voting system based on the number of detectors in a space is integrated into the safety IAS, allowing the operator to know if there is a confirmed fire or gas release for that section of the vessel, after which a set of actions are automatically triggered without operator intervention. For example, if two smoke detectors are triggered in the laundry room, it will be considered a confirmed fire. Once this has occurred, ventilation and power to the space could automatically be cut off and the general alarm sounded. Depending on the nature of work being done in the space and the voting system, there could be a delay prior to any of this being set off, or steps could be taken by the operator to have the ability of setting off these actions only after being notified by the system. The same thought process goes for hydrocarbon gas or hydrogen sulfide gas releases. This monitoring system, with "cause and effect" logic, is provided for every space aboard the vessel. The cause-and-effect page for each space shows the operators what will be triggered (effect) on the basis of the alarms received (cause).

Detector in field

The Safety IAS also provides a means for monitoring and activating firefighting systems, such as

- main fire pumps
- jockey pumps
- Inergen system (CO_2 activation will not be possible from a safety IAS since it is lethal)
- deluge system
- galley exhaust ducts fire systems (such as CO_2 or Inergen)
- helideck foam system
- water mist system

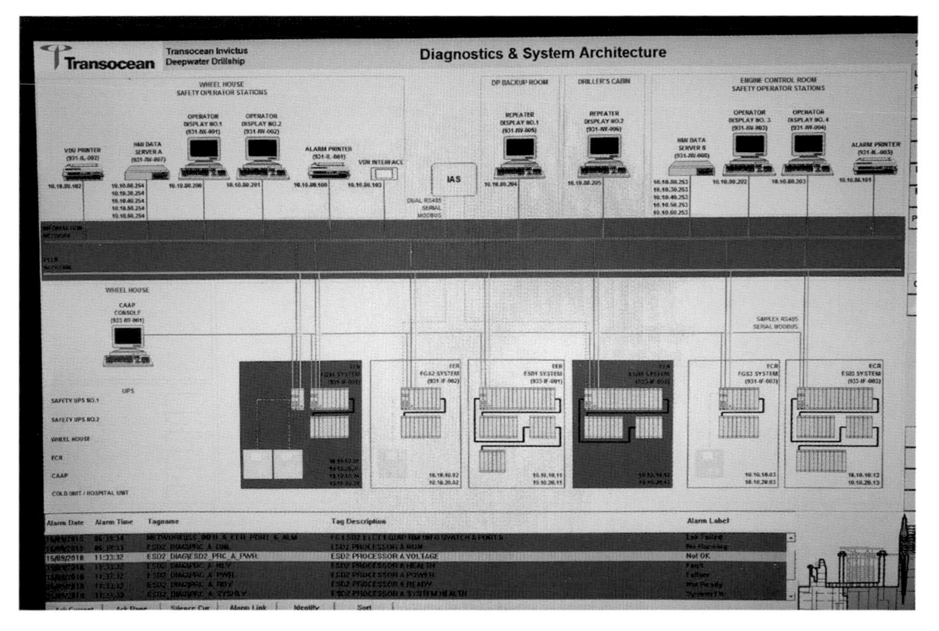

Overview of the Safety IAS system. In the picture above, there is an alarm on the system highlighted in red.

2.11.2. Emergency Shutdown

Emergency shutdown (ESD) systems are used for shutting down all equipment related to a particular area (i.e., all power being supplied to that area, any equipment in that area, and ventilation [fans and dampers]). ESDs are classed by various areas being shut down, with the highest level of ESD being a total shutdown of the entire vessel, including tripping breakers for backup batteries that might be running smaller equipment (this will be used only during emergency evacuation of the vessel).

The main control panel for the ESD will be on the bridge and has the capability to shut down the entire vessel by compartment. The rig floor will have an ESD panel to secure only the drilling module (rig floor equipment and mud module equipment), while the engine room will have an ESD panel for shutting down various engine rooms. The following is the layout and nomenclature of an exemplary ESD system:

ESD 0 (drilling shutdown)
ESD1A (port engine room shutdown)
ESD1B (center engine room shutdown)
ESD 1C (starboard engine room shutdown)
ESD 2 (emergency power shutdown)
ESD3 (abandon UPS shutdown)

ESD activation point for ESD3 will also be available adjacent to the lifeboats and the helideck, but for these to function, a switch on the bridge will need to be activated.

ESD on bridge

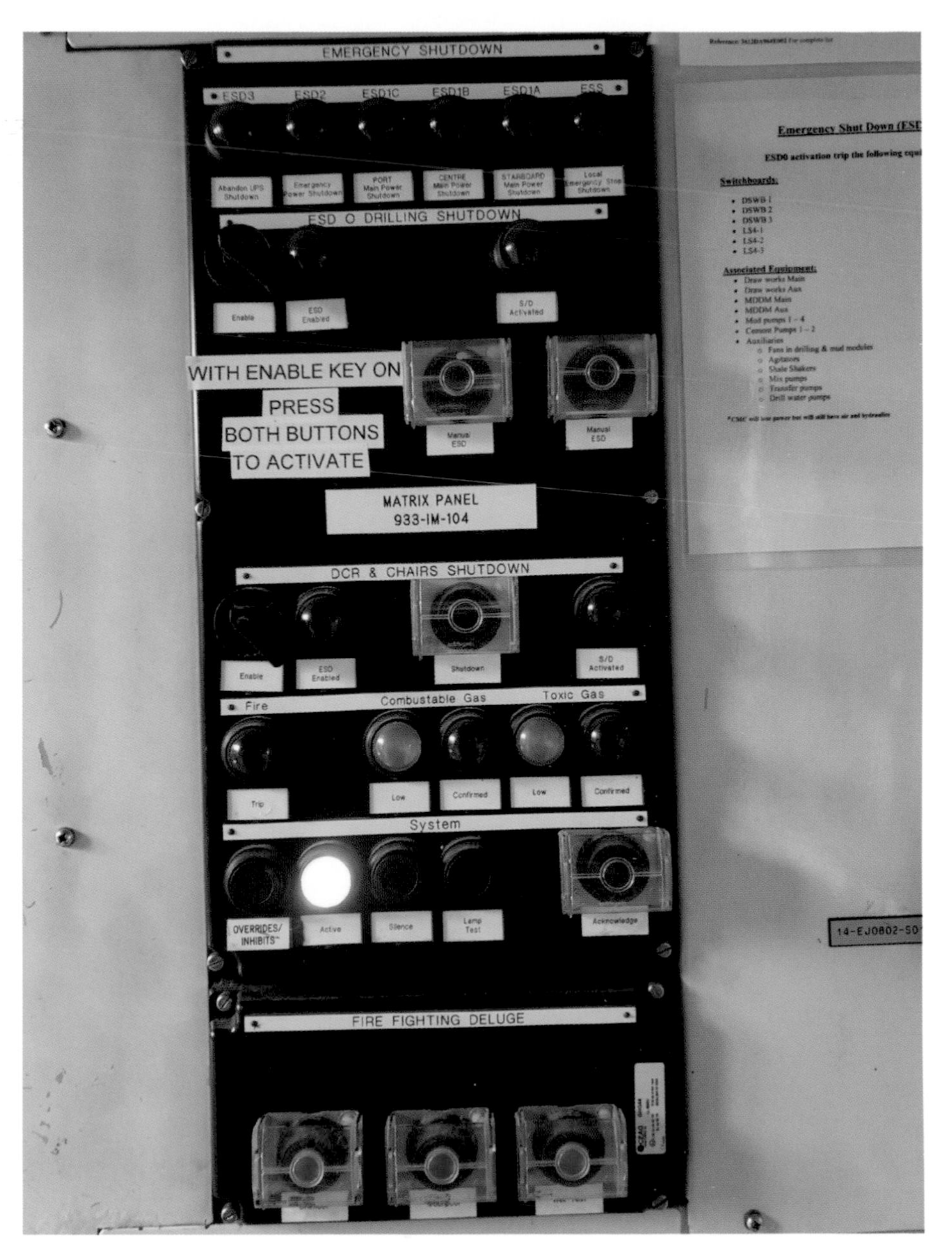

ESD button on rig floor

2.11.3. Critical Panel

The critical alarm panel gives the operator a general overview and status of all the areas on board the vessel. On the basis of the philosophy in the safety IAS, the operator will know if there is a confirmed fire or confirmed gas in a space. In addition to just status view, the critical alarm and action panel gives the operator the ability to start fire pumps and jockey pumps and to shut down ventilation in general areas of the vessel. In the picture, the forward machinery space, aft machinery space, drilling area, and accommodation are areas where ventilation can be shut down by a single push button. What is important to note is that the critical alarm and action panel are on a separate network from the safety IAS, providing an additional layer of activating certain systems if both networks on the safety IAS were to go down.

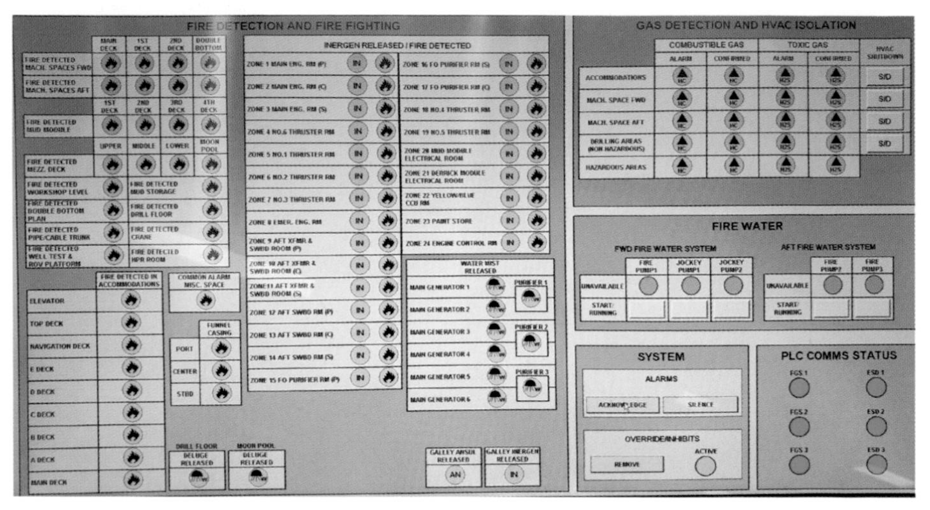

CAAP Panel

2.12. Vessel Management System (VMS) / Integrated Automation System (IAS)

VMS—sometimes referred to as the Integrated Automation System (IAS)—provides the interface between a lot of equipment on board and the operator, either on the bridge or in the engine control room (adjacent to the main engines aft of the vessel). From the VMS the operator is able to function or monitor the following minimum systems:

- ballast system
- bilge system
- ventilation (dampers and fans)
- salt water for auxiliary systems and main engines.
- fuel oil system
- main diesel generators (MDG) and power distribution
- thrusters alongside its auxiliary systems
- main rig air system and start air for MDGs
- drilling-fluid tanks (see p. 56, 3.3 Reserve pits and Active Pits)

There are typically three VMS stations in the bridge, three in the engine control room, and one on the rig floor. The ability to take control of certain equipment is possible from various locations, depending on the authority level selected by the vessel owner for the personnel in that space.

VMS systems operate equipment all over the vessel; thus each VMS station is connected to field stations all over the vessel. This is done through dual redundant ethernet connection. When the operator of the VMS sends a signal, it goes to the field station, and from the field station the signal is sent to the specific equipment it controls. For example, when an operator sends a signal to stop a fan on deck, the signal travels from the VMS to the field station that controls the fan. The field station in turn sends that signal to the fan, and once stopped the close signal is sent back to the field station and in turn to the VMS, closing the loop and letting the operator know the fan has been shut off. This is the operation for the majority of the equipment controlled by the VMS.

Example of field station

Within the VMS is a built-in power management system (PMS). The main role of the PMS is to prevent the vessel from completely losing all power–essentially having no engines online, which is called a blackout. This is accomplished by having certain settings in place to never have the engines run at full capacity to the extent where they can potentially trip from overspeeding; multiple other settings can be put in place to prevent a blackout.

The basic setting monitors how much load is on the main bus, and once the load reaches a preset limit for a certain period (for example, 85% load on one engine for one minute), an additional generator is automatically connected to the main bus. If all available generators are online and the load on the bus stays at a certain limit, the system can start reducing power output to noncritical equipment to maintain a certain reserve; typically drilling equipment first, since it consumes a lot of power. If this does not accomplish the task of reducing the load on the main bus, the thrusters will also have a reduction on how much power is available to them. These two items are the main consumers on board drill ships. Modern drillships typically have six thrusters, which can consume as much as 5.5 mW at full capacity each; having all at full capacity will consume 33 mW. With six generators at full capacity producing 6.75 mW, each produces 40.5 mW, so most modern vessels have sufficient power to run all thrusters at full capacity without coming close to potentially causing a blackout.

ECR with operator stations

Module Parameters - PS039: HS3

Page 1 | Page 2 | Page 3

MODULE TYPE : swbd

ASYMMETRIC LOADSHARING

Asymmetric main load	80.0	%
Asymmetric main period	4.0	hours
Asymmetric top-up minimum	25.0	%

LOAD DEPENDENT START/STOP AND CONSUMER CONTROL

Lock advice mode (0/1/2-unl/on/off)	0	
Lock cons control (0/1/2-unl/on/off)	1	
Enable load dep. start select (0/1/2)	1	
Start limit type (%/kW)	0	
Enable load dep. stop select (0/1/2)	2	
Stop limit type (%/kW)	0	
Start limit 1 delay	30.0	sec
Start limit 2 delay	10.0	sec
Stop limit delay	900.0	sec

START/STOP LIMITS :

Connected DDGs	Start 1 %	Start 1 kW	Start 2 %	Start 2 kW	Stop %	Stop kW
1	72.0	0.0	85.0	0.0	0.0	3000.0
2	75.0	0.0	85.0	0.0	20.0	3000.0
3	78.0	0.0	85.0	0.0	30.0	3000.0
4	80.0	0.0	85.0	0.0	40.0	3000.0
5	87.0	0.0	90.0	0.0	40.0	3000.0
6	100.0	0.0	100.0	0.0	45.0	3000.0
7	100.0	0.0	100.0	0.0	0.0	3000.0
>7	100.0	0.0	100.0	0.0	0.0	3000.0

Master | PS A | PS B | Print | Close | Help

Engine Start & stop settings

2.13. DP System

Bridge layout

The Dynamic Positioning (DP) system is designed to automatically maintain the vessel's position and heading as set by the dynamic-positioning operator (DPO). The system receives multiple inputs from the vessel management system, reference sensors (GPS satellites, acoustics on the seabed), motion sensors (calculating the vessel's roll, pitch, and heave), gyro units (for the ship's heading), and wind sensors and computes what thrust is needed from each thruster to maintain the vessel's position and heading. The DP system then automatically sets a certain amount of thrust to each of the thrusters to maintain required position and heading; the thruster forces are actively updated as the vessel moves. This position and required heading are set by the DPO. In addition to a set position and heading, the operator can move the vessel at a required speed to a required location, making for a very controlled move to and from an area. Systems also have the capability of inputting data coordinates and allowing the vessel to essentially maneuver from a certain area, turning at a predetermined set point until the desired location is reached.

DP systems on a modern drillship have three operator stations on the bridge and at least one operator station either in the engine control room or an area at least separated from the bridge by an A-60 bulkhead.

Classification societies have several redundancy levels, and the modern drillship is generally a DP-3 or DYNPOS-AUTRO vessel. The redundancy goes further than the location and number of sensors and DP control systems; as depicted in the table, the power distribution and even cable supplying certain systems are required to be physically separated (also an A-60 separation). This goes beyond the scope of this book, but the separation is critical to the design of the vessel.

A misconception that has plagued the industry (a lot of DP operators, company representatives, and even managers) is the change of vessel status from DP3 to DP2. It has been customary for people to believe that if they lose (through some temporary failure) one reference sensor (which is in your alternate area as per the table), the vessel now goes from DP3 to DP2. This is completely wrong, since the vessel's class notation is based on the design of the vessel and not the current operating parameters. If the DPO decides to change certain parameters for whatever reason, you can not say that the vessel if no longer a DP3 vessel; this is misleading. The status of a vessel (DP3, DP2, or DP1) is based off the design and construction of the vessel, so changing a setting on the operating system does not change your DP-class status.

Subsystem or component			*Minimum requirements for class notations*			
			DYNPOS-AUTS	DYNPOS-AUT	DYNPOS-AUTR	DYNPOS-AUTRO
			DPS 0	DPS 1	DPS 2	DPS 3
Electrical power system	Electrical system		No-redundancy 3)	No-redundancy 3)	Redundancy in technical design	Redundancy in technical design and physical separation (separate compartments)
	Main switchboard		1 3)	1 3)	1	2 in separate compartments
	Bus-tie breaker		0 3)	0 3)	1	2, 1 breaker in each MSB
	Distribution system		Non-redundant 3)	Non-redundant 3)	Redundant	Redundant, through separate compartments
	Power management		No	No	AUTR: Yes DPS 2: No	AUTRO: Yes DPS 3: No
Thrusters	Arrangement of thrusters		No-redundancy	No-redundancy	Redundancy in technical design 4)	Redundancy in technical design and physical separation (separate compartments)
	Single levers for each thruster at main DP-control centre		Yes	Yes	Yes	Yes
Positioning control system	Automatic control; number of computer systems		1	1	2	2 + 1 in alternate control centre
	Manual control; independent joystick system with automatic heading control 2)		No	Yes	Yes	Yes
Sensors	Position reference systems		1	2	3	3 whereof 1 in alternate control centre
	External sensors	Wind	1	1	2	2 whereof 1 in alternate control centre
		Gyro compass	1	1	3 1)	3 1) whereof 1 in alternate control centre
		Vertical reference sensor (VRS)	1	1	AUTR: 3 DPS 2: 2 5)	3 whereof 1 in emergency control centre
UPS			0	1	2	2 + 1 in separate compartment
Printer			Yes	Yes	Yes	Yes
Alternate control centre for dynamic positioning control back-up unit			No	No	No	Yes

1) One of the three required gyros may be replaced by a heading device based upon another principle, as long as this heading device is type approved as a TDH (Transmitting Heading Device) as specified in IMO Res. MSC.116 (73). For notation **DYNPOS-AUTRO** and **DPS 3** this is not to be the gyro placed in the alternate control centre.

2) The heading input may be taken from any of the required gyro compasses.

3) When this is part of the ship normal electrical power system (i.e. used for normal ship systems, not only the DP system), then Pt.4 Ch.8 applies.

4) For **DPS 2** see also B202.

5) Where necessary for the correct functioning of position reference systems, at least three vertical reference sensors are to be provided for notation **DPS 2**. If the DP-control system can position the ship within the operating limits without VRS corrections, only 2 VRSs are required.

Guidelines for DP system

3 DRILLING EQUIPMENT

Subsea BOP

Moonpool with riser

This section discusses what components of a deepwater drilling rig make up the drilling equipment. To do so, it is important to explain the path that drilling fluid (mud) takes during normal drilling operations. It is further important to note that the drillship is connected to a well (on the seafloor) by means of the riser and blowout preventer (BOP). The riser is connected to the drillship, which in turn is connected to the BOP, and hence the well. It is through this medium (riser) that the drill pipe (and hence the drilling mud) makes its way to the formation for drilling operations.

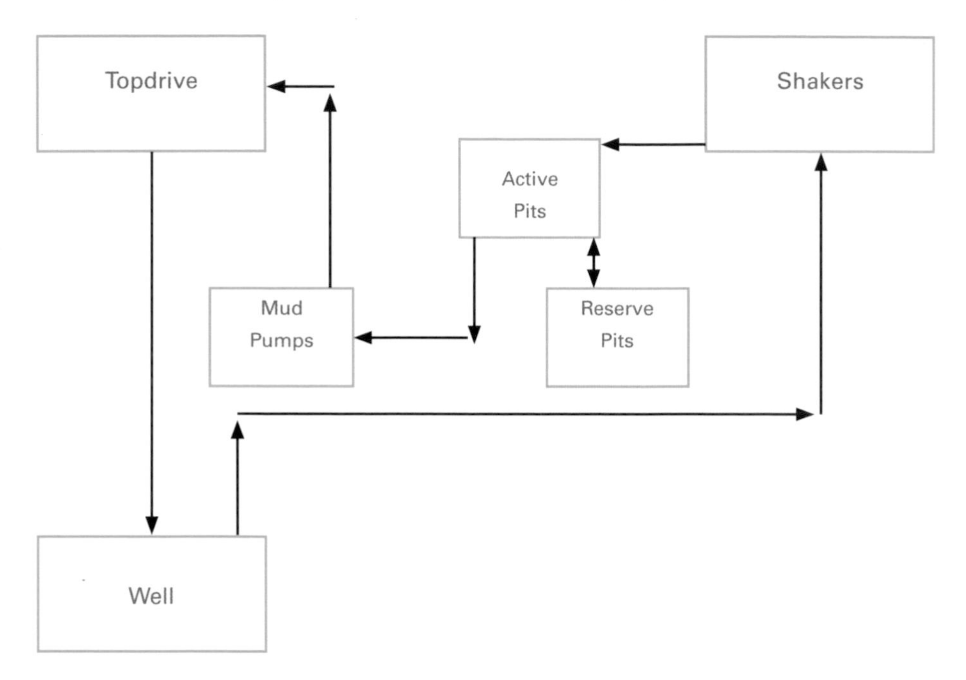

3.1. Mud Cycle

The line diagram depicts a simplistic flowchart of the typical mud path during regular drilling operations. Each component of the system will be discussed in this chapter.

Mud is stored in the reserve pits (holding tanks) and transferred to the active pits by using mud transfer pumps. In the active pits, mud is processed as required, and with the use of mud pumps the mud is pumped to the top drive after having gone through the standpipe manifold and through the standpipe. From the top drive, mud enters the drill pipe through a swivel and into the well. Mud returns to the vessel from the well through the annulus (space between the riser and the drill pipe), and once it reaches the surface (vessel) it goes to the shale shakers (shakers). From the shakers, mud is sent back into the active pits to complete the cycle.

3.2. Mud Pumps

The mud pump is the primary component in moving fluids during normal drilling operations. On modern drillships, mud pumps are single-acting triplex pumps. A single-acting triplex pump is a pump that has three pistons moving inside three liners. In a single-acting pump (as opposed to double-acting or duplex pump), the pistons move back and take in mud through open intake valves, and then the pistons move forward and push the mud through discharge valves. Double-acting pumps move mud during both strokes of the piston and hence can move more volume; for the following reasons, triplex pumps are preferred over duplex pumps.

Triplex pumps weigh 30% less than a duplex of equal capacity. The lighter-weight parts are easier to handle and therefore are easier to maintain. They cost less to operate, their fluid end is more accessible (again making them easier for maintenance), and they discharge mud more smoothly—that is, a triplex's mud output doesn't surge as much as a duplex's. One of the most important advantages of a triplex over a duplex pump is that they can move large volumes of mud at the high pressures

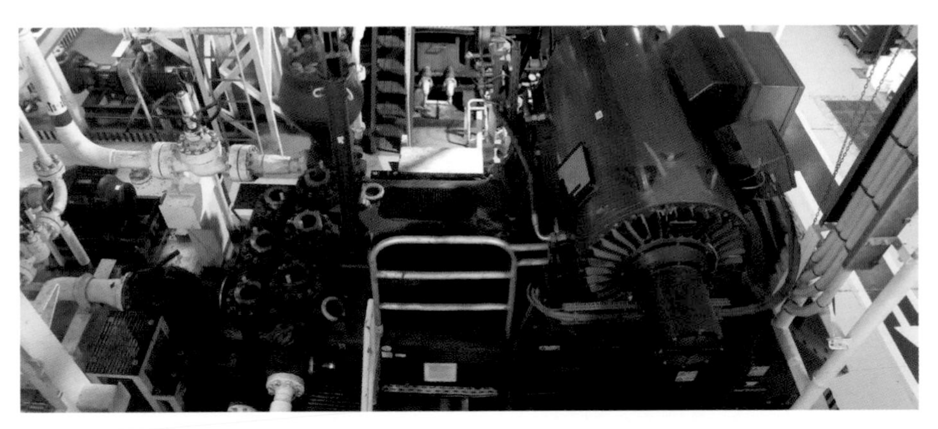

Overview of triplex mud pump

Topview of mud pump

Bottom view of mud pump

required for modern deep-hole drilling. High pressure is essential for deepwater drilling, and this key factor favors triplex pumps over duplex pumps, since they provide as much as 7,500 psi of discharge pressure.

Both on the intake and discharge side of mud pumps, pulsation dampeners are installed. On the intake side, this pulsation dampener prevents surges from the pump's suction and, on the discharge end, dampen the outgoing flow to the standpipe. This reduction in surge on the discharge side also reduces the "noise" made by the pumps as the mud is pumped down the hole. This is very important, since special tools are used to communicate with the well (data transfer through the mud), and excessive noise and surges in the flow will prevent these tools from operating properly.

3.3. Reserve Pits and Active Pits

Reserve pits are tanks typically at the lowest level of the vessel. This is where drilling mud that is not actively being used is stored either for future use or to later be discharged to make room for a different type/composition of mud. There are anywhere between six and nine reserve tanks on drillships, and each can hold approximately 10,000–12,000 bbls. of mud. To maintain proper homogeneity of mud in these tanks, agitators are used. Agitators are mounted at the top of the tank and contain a shaft that drives an impeller, which in turn mixes the mud.

Active pits are very similar to reserve pits, except for the fact that the individual tanks are a lot smaller; as such, it is not uncommon for a drillship to have more than twelve active pits. In addition to the size, active pits are in the same space or a compartment adjacent to the mud pumps, and as such are either on the main deck of the vessel or a few levels above the main deck. Mud is transferred to and from the active and reserve pits through mud transfer pumps. Active pits also contain agitators similar to those in the reserve pits.

Reserve pit room

Agitators

3.4. Top-Drive and Hoisting Equipment

Before talking about the top drive, there are a few items to mention that work along with the top drive and the path the mud follows prior to getting into the top drive.

From the mud pumps, the mud goes through the standpipe. The standpipe is a fixed piping between the mud pumps to the top of the derrick. Lineups from the mud pump allow for the mud to flow to at least two standpipes (*illustrated*) and the booster line (see p. 62).There are two standpipes because dual-activity rigs have two top drives, and depending on which top drive is being used (main or auxiliary), you will need to line up either to standpipe #1 or standpipe #2. Furthermore, certain drillships allow for a crossover valve on the derrick so that either standpipe can be used for both top drives. From the top of the derrick, a flexible but rigid hose (called a kelly hose) connects from the standpipe to a connection on the top drive.

Standpipe manifold from mud pump room

Standpipe manifold connection to kelly hose

3.4.1. Hoisting Equipment

Hoisting equipment suspends the drill string in the hole. It also allows the driller to raise and lower the drill string into and out of the hole. Further, it allows the driller to adjust the weight on the bit, which is required to make the bit drill. The hoisting equipment consists of

- the crown block
- the traveling block and hook
- the drilling line
- the drilling line supply reel
- the deadline to crown block
- the fast line to draw works
- the draw works
- the deadline anchor

Rig picture with hoisting system

The draw works is controlled by the driller from the drill shack and allows him to lower or raise the traveling block. The crown block is fixed at the top of the derrick, while the traveling block "travels" up and down the length of the derrick. It is from the traveling block that the top drive is secured through the drilling hook.

Topdrive

The top drive, also known as the MDDM (Modular Derrick Drilling Machine from a specific vendor), provides the torque necessary to rotate the entire drill string and hence is the backbone to the drilling operation. The MDDM contains several parts:

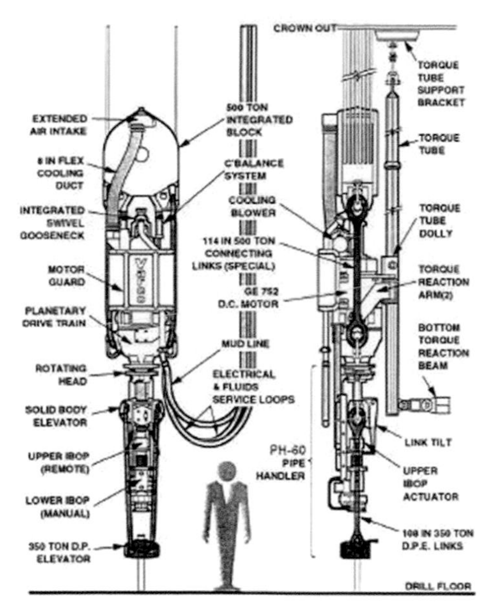

IDS-1 General Arrangement & Components

Top drive with labels

- IBOP (inside blowout preventer)
- elevator links and elevator
- service loops
- link tilt
- drive system
- weight-compensating system
- retractable dolly

Drive system:

The main component of the top drive is the drive system. The drive system consists of AC motors (typically at least two for redundancy) whose torque is transmitted to a swivel that connects to the drill pipe through a saver sub and rotates the drill string for drilling the well. The saver sub provides the crossover between the top drive and the drill pipe. Varying sizes of saver subs allow the driller to connect to different-sized drill pipes. In addition, the saver sub protects the IBOP.

IBOP:

The IBOP is a remotely controlled valve that allows the driller to control the flow of mud into the drill string. The IBOP also allows mud to flow only one direction, from the top drive to the drill string and not the other way around. Modern top drives typically have three IBOPs within the top drive.

Service Loops:
The service loops are made up of several lines consisting of

- power cable supplying the electrical power necessary to run the motors
- hydraulic lines necessary for the operation of hydraulically actuated valves on the top drive
- freshwater cooling lines
- air lines from the main rig air system also used for cooling or activation of equipment

Weight-Compensating System:
During drilling operations, when one joint of drill pipe is connected and disconnected from another, the weight-compensating system prevents too much weight from being placed on the threads. For example, when disconnecting drill pipe, once the tool joint has been disconnected the weight-compensating system gradually and slightly lifts the disconnected pipe from the remainder of the drill string.

Elevator and elevator links

Elevator Links and Elevator:
Elevator links (two—one on each side of the top drive) support the elevator on one end, and the other end is secured to the top drive. Different sizes of elevator links and elevators are used depending on what size of drill pipe is being used. Elevators (which can be hydraulically or pneumatically activated in most cases) are hinged mechanisms that may be closed around the drill pipe or other drill string components to facilitate lowering them into the well bore or lifting them out of the well bore. In the closed position, the elevator arms are latched together to form a load-bearing ring around the component. In the open position, the device splits roughly into two halves and may be swung away from the drill string component; this is where the pipe is lifted from.[1]

[1] http://www.glossary.oilfield.slb.com/Terms/e/elevator.aspx

LinkTilt:
Link tilt assists the elevator links in tilting either toward or away from the well center. Once a string of drill pipe has been lowered into the hole and is secured in the well center by means of slips, the elevators open up and the link tilt pushes the elevator away from the drill pipe, allowing the top drive to be retracted up the derrick to get the next string of pipe.

Retractable Dolly:
The retractable dolly is designed to retract the entire top drive away from the well center when running up and down the derrick. One end is made fast to a vertical track that runs the entire length of derrick, and the other end is secured to the top drive.

3.5. Drill Pipe and Pipe Handling

Drill pipe, prior to being used, is stored horizontally on the pipe deck, which could be either forward or aft of the drill floor. For proper preservation and stowage, timber is laid onto the deck and between each layer of drill pipe. For ease of nomenclature, drill pipe is referred to by the weight per foot of a particular pipe. As such, "27 pound" drill pipe weighs 27 lbs./ft., and so is 40, 50, 65, etc.–and they all typically measure between 27 and 32 ft. long. Smaller-length pipes are available, called pup joints, and to change over from one size pipe to another, "crossovers" or "subs" are used, which can be thought of as adapters.

Pipe deck with pipe stored

The main equipment used for handling drill pipe on the rig floor includes forward and aft conveyor belts and the drill floor manipulator arm (DFMA).

Forward Conveyor Belt:
From the pipe deck, drill pipe (and also equipment transported from the pipe deck to the rig floor and vice versa) is loaded onto the forward conveyor by using cranes. The conveyor has a skate that moves the pipe (still in a horizontal position) onto the rig floor, where it is picked up by the elevators, and the top drive moves the drill pipe to a suitable vertical height, where it is then transferred to the vertical pipe-handling system.

Pipe being moved from forward conveyor to rig floor

3.6. Vertical Pipe-Handling System

The vertical pipe-handling (VPH) system constitutes the bridge crane (BRC) and the lower guiding arm (LGA). The bridge crane is above the fingerboard (where pipe is stowed in the vertical position), and the lower guiding arm has a track on the rig floor, allowing it to run forward and aft of the rig floor. Once the drill pipe has reached the height necessary for stowage (by use of the top drive), the weight of the entire drill pipe is transferred from the top drive to the bridge crane, while the LGA guides the drill pipe to a suitable position. Once in the stowage position in the fingerboard, the bridge crane lowers the drill pipe on deck (still in a vertical position) and "finger-latches" on the fingerboard prevent the drill pipe from moving.

LGA guiding pipe

BRC with fingerboard

Iron Roughneck:
The iron roughneck is a hydraulically operated machine used to connect or disconnect drill pipe from the well center. It can be brought to the well center and retracted remotely from the drill shack. How much torque is needed to make connections is also preset by the crew, along with the size of drill pipe that will be used to make connections.

Iron roughneck retracted

Iron roughneck with LGA

Drill Floor Manipulator Arm (DFMA):
The DFMA is used to guide various sizes of drill pipe or equipment around the rig floor from the conveyor area to the setback area, and even the well center. Its primary purpose is to guide equipment and prevent it from swinging around the rig floor.

3.7. Annulus and Diverter

Annulus:
From the top drive, the mud is pumped into the well through the drill pipe. At the bottom of the drill pipe, the mud exits through the drill bit and back up toward the rig. The space that the mud travels in at this stage (space between the drill pipe and the borehole or casing) is called the annulus. The mud brings with it cuttings all the way to the surface and gets to the diverter.

Drill bit on deck

Drill bit on deck

Diverter Housing:
Once the flow of mud comes off the drill bit and returns up the annulus all the way to the vessel, it comes into contact with the diverter.

The diverter helps in "diverting" returns from the well either directly to the shakers (during normal operations) or overboard in case of uncontrolled flow of mud to the surface. Depending on which side the wind is blowing, drillships are able to divert the flow overboard to either the port or starboard side. During certain operations, drilling fluids are diverted to the trip tanks from the diverter as opposed to the shale shakers.

Diverter being handled on rig floor

Diverter going port and starboard

3.8. Surface Mud Processing

3.8.1. Shale Shakers

Top view of shale shakers

Shale shakers are a set of vibrating screens used to remove cuttings from the mud that returns to the surface while drilling is ongoing. When mud goes through the diverter, under normal drilling operations it goes through a flow diverter (known as the "gumbo box") and from here is sent to the shakers. As the name implies, the flow diverter allows crew members to flow the mud to various areas. From the flow diverter, mud is sent to the shale shakers, where the cuttings are captured by screens while the mud falls into mud tanks. Screens are also changed depending on the size of returns that need to be captured and the running hours of a particular shaker. Offshore, the cuttings are stored in tanks and sent back ashore either for proper disposal or further analysis. It is important to note that the shakers are typically the first place on board the vessel where the mud is exposed to the surface (atmosphere) and, as such, are in an area where the crew maintain a close watch on how fast returns are coming in, the size of returns, and the gas levels in the shaker house.

Modern drillships typically have eight shakers in the shaker house, arranged in parallel. Mud flows from the diverter into the "gumbo box," and valves are opened to direct the flow of mud either to one or multiple shale shakers. The number of shakers used depends on how fast the rig is getting returns and how many shakers can keep up.

Shale shaker with screen

3.8.2. Desander and Desilter

After the cuttings are trapped by the shakers, the returning mud drops into a mud return tank or sand trap tanks (or both). These tanks contain bulkheads that are only about 70%–80% the height of the actual tank, and return mud overflows from one compartment to the other, allowing for residue to settle at the bottom of these tanks; crew members occasionally have to clean out this residue. If the mud is still not cleared from impurities after this process, the mud is sent into a desander/desilter tank, which sends mud through the desander or a desilter, or both. A desander is a centrifugal device that allows sand to settle on the sides of a cone as mud is pumped through it, thereby removing sand; a desilter operates similarly to desanders but will remove even-finer impurities. Because of the depth at which deepwater drillships operate, desanders/desilters are hardly utilized in drillship operations, but this equipment is present.

3.8.3. Mud/Gas Separator and Vacuum Degasser

When mud returns from the well, it sometimes contains some gas that has been absorbed by the mud. The gas levels (which can be either toxic, such as H_2S, or explosive hydrocarbon gas) contained in the mud are actively tracked, and if gas levels start rising, the mud is routed through the vacuum degasser before going to the shakers. As the name suggests, the vacuum degasser is a tank that creates a vacuum, thereby extracting any gas (hence the name "de-gasser") in the mud. The extracted gas is vented at the top of the derrick, and the mud is sent back to the mud tanks.

When the vessel is circulating out a kick (influx of gas from the formation) or is under a well control situation and mud is circulated through the choke and kill lines (see p.86), rather than going directly to the shakers, mud is sent to the mud/gas separator (MGS). The MGS tank is bigger than the vacuum degasser and contains baffles that are angled; the mud spills over from one set of baffles to the next, thereby allowing the mud to be spread out (increasing the exposed surface area), and any gas in the mud escapes through a vent that, similar to the vacuum degasser, ends up at the top of the derrick. From the MGS, mud can be sent back toward the flow diverter, where it goes through the processing detailed earlier.

Vacuum degasser

3.9. Trip Tanks

A trip tank is a specialized mud tank used to accurately measure how much mud is pumped into or out of the hole, to ensure the rig does not have more or less mud than what is required when tripping. When pulling drill pipe out of the hole, the volume of pipe pulled out has to be replaced by mud of equal volume; this is done using a trip tank. Through this method, crew members accurately know how much mud has been pumped into the hole. Drillships typically have two trip tanks ranging between 60 and 120 bbls. Trip tanks are lined up to either go to the flow divider or the diverter housing.

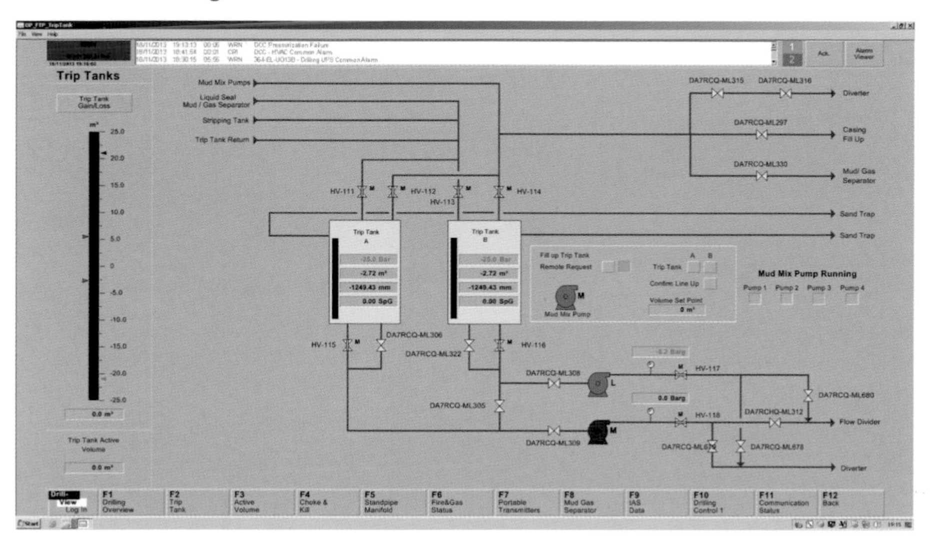

Overview of mud path

3.10. Bulk System and Fluid Tanks

The bulk system is used to classify three main components used in drilling. Barite and bentonite are used for weighing up drilling mud (making it heavier), while cement is used to keep the casing in place and thereby isolate certain zones while drilling. The cement unit is typically run by the cementer, while barite and bentonite and its additive are mixed by crew members, with directives coming from a mud engineer.

All substances arrive at the drillship from offshore supply boats and are transferred through a hose connecting the supply boat to the drillship. Barite, bentonite, and cement all are stored in tanks below the main deck in the "bulk tank room," which usually contains at least eight bulk tanks each with a capacity of about 85–100 m³. It is important to note that barite/bentonite usually use the same piping (but different tanks), while the cement system has separate piping. Crossover valves typically exist if an additional tank is to be used from either side, but this hardly ever happens, since tanks will need to be properly cleaned to avoid any contamination.

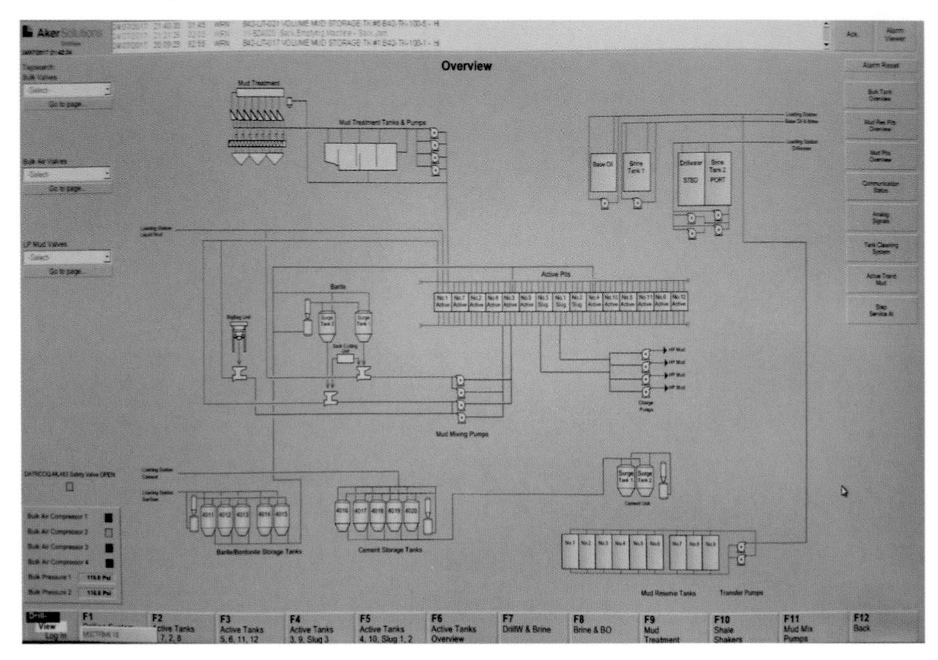

Bulk tank system

3.10.1. Barite/Bentonite System

Barite and bentonite are principally used for weighing up drilling mud. Different sections of drilling require mud to have a certain weight, and as such, going from a mud weight of 12.5 ppg (pounds per gallon) to 13.0 ppg, the principal additive is either barite or bentonite, or a combination of both, along with other additives determined by the mud engineer. How much of each needs to be added is decided by the mud engineer, and through their directive, crew members are primarily involved in weighing this up. Rig air is pumped into these bulk tanks to provide a certain amount of pressure. Through a series of valves, the bulk is then transferred from the main bulk tank to the surge tanks (smaller tanks), where it is used to weigh up the mud. The surge tanks are typically in a room adjacent to the active pit room, called the sack storeroom.

Mud is always weighed up from the active pit tank. To do this, a set of mud/mix transfer pumps are used to circulate mud from one active tank to another. During the transfer of mud from one active tank to another, the mud is circulated through the mud/mix hopper. It is from the mud/mix hopper that barite/bentonite is dumped into this line and thereby weighs up the mud. Once the mud arrives in the new active tank, the mud engineer takes samples and either recirculates to get the right dosage or continues with the process. It is important to note that in the hopper, there are other additives that go into the mud, which goes beyond the scope of this book. The main purpose of the hopper is to provide safe access to the mud engineer for placing additives into the drilling mud.

Bulk tank room

3.10.2. Cement System

The cement system is very similar to the barite/bentonite system, with the exception that its surge tanks are used by the cementer to prepare cement slurry, which is pumped downhole during cementing operations. The cement unit has its own pumps and a separate pipe that goes to the rig floor; this line is usually connected to the choke/kill manifold (see p.92) or could very well be an independent pipe. From the independent pipe, crew members connect hoses to the drill pipe, and in this way cement slurry is pumped downhole, using the pumps from the cement unit.

Cement room

Surge tanks

The cement unit is also used when pressure-testing equipment on the rig floor, surface equipment used for well control (such as the choke/kill manifold), a variety of equipment on the BOP, and a multitude of other well control tools. The main reason why the cement unit is used is because smaller volumes (and more-accurate amounts) can be pumped from the cement unit to this equipment and high pressures can be attained, so smaller volumes at accurate pressures are the main reason why the cement unit is used for pressure-testing most equipment used during drilling operations.

3.10.3. Fluids

3.10.3.1. Drill Water

Drill water is usually generated on board or taken from offshore supply boats and is typically transferred from the potable water tanks to the drill water tanks. Offshore supply boats also transfer water used as drill water. The water makers that generate potable water hence have a dual purpose: they are used for generating both potable water and drill water. Drill water is then transferred into the active pit tank, and the mud engineer uses this for mixing his mud. There are usually two drill water tanks, and they are usually in the same vicinity as the reserve pits, with two drill water transfer pumps also in the same area/vicinity. These are regular tanks with a total capacity ranging from 15,000 to 20,000 bbls. In addition, drill water piping generally goes around the entire vessel for general use, such as items that need to be cleaned with fresh water as opposed to salt water. This is very similar to the general saltwater piping covered in the "Marine Equipment" section.

3.10.3.2. Base Oil

Base oil is another component added into drilling mud during drilling operations. Base oil, just like drill water, has its own tank and its own pump (base oil pump). Piping from the base oil tank allows for the base oil to be transferred to the active pit tanks and sometimes the cement unit as well. It is taken on board the vessel just like all other fluids via transfer from an offshore supply boat.

3.10.3.3. Brine

Brine is usually used only in completion operations/phase of a drilling program and when entering a pay zone. They can range in weight from 8.4 to 20 ppg and consist mainly of inorganic salts. Because they contain no heavy particles that can potentially plug or damage the formation, their primary role is in well completion. Just like with the previous two tanks, drillships are also equipped with a brine tank and an independent pump.

It is typical for tanks containing these fluids (drill water, base oil, brine, and reserve pits) to be in the same general location. This allows for one tank to be used to store additional fluids that will not be used for a prolonged period, and additionally, crossover valves allow for this to be used. For example, if during a drilling campaign the customer will not be using brine, that tank can be used for storing additional drill water. Care must be taken when this is done, and proper tank-cleaning procedures must be used if said tank is to be used for its original use.

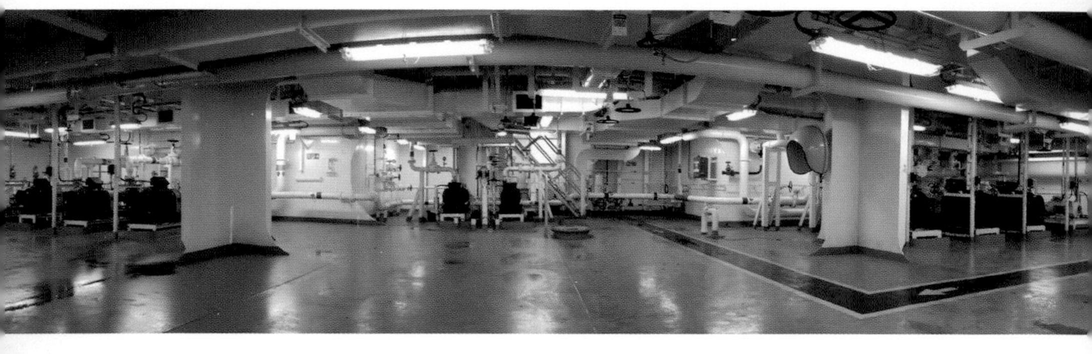

Mud transfer pump room

4 SUBSEA EQUIPMENT

4.1. BOP and LMRP

The blowout preventer (BOP) is a set of large, specialized valves designed to control and monitor the flow of oil or gas (or both) to prevent a blowout (hence the term "blowout preventer"). One can therefore imply that a blowout is an uncontrolled flow of oil or gas (or both) from a well. The discussion begins with what surface BOPs are made up of, and from there will dive into BOPs used in deepwater drilling wells (also called multiplex [MUX] / subsea BOP). It is important to note that a BOP is installed only once the well being drilled has reached a certain depth and a certain size of casing has been installed in the well; this depends on the expected formation pressures at various depths. Once installed, all drill pipe and fluids go through the marine riser and BOP when coming from, or going to, the surface.

4.1.1. Surface BOP

On land-based rigs, jack-ups, and certain midwater rigs, the BOP is on the surface (not below the water). Surface BOPs are less complex but accomplish the same task as a subsea BOP. Primary parts of a BOP include

- annular preventers
- blind rams
- pipe rams and variable-bore rams
- shear rams
- accumulator
- choke/kill manifold
- BOP panel

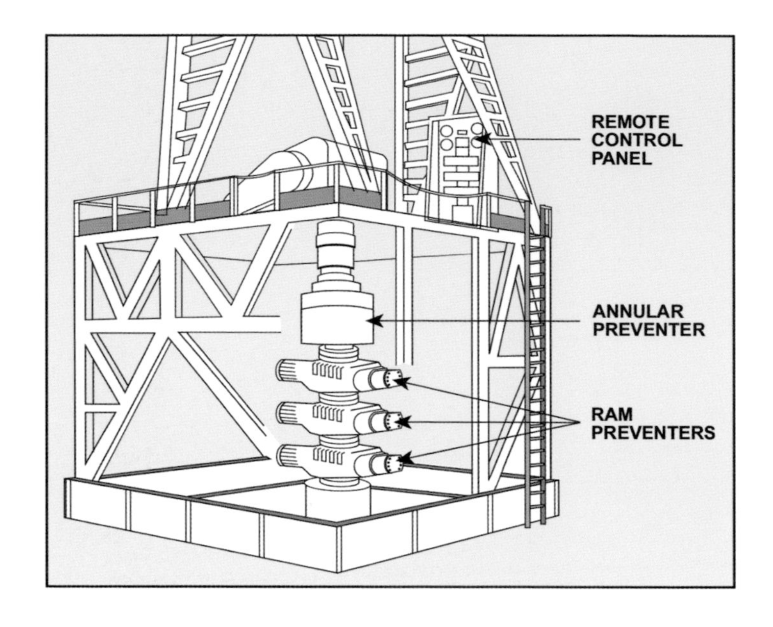

Surface BOP

Annular Preventers:
Annular preventers are at the topmost section of the BOP. Typically the first set of rams to be closed when a well control event occurs, it is made up of synthetic rubber that provides a seal around any size of pipe, thereby creating a barrier between the well and the rig.

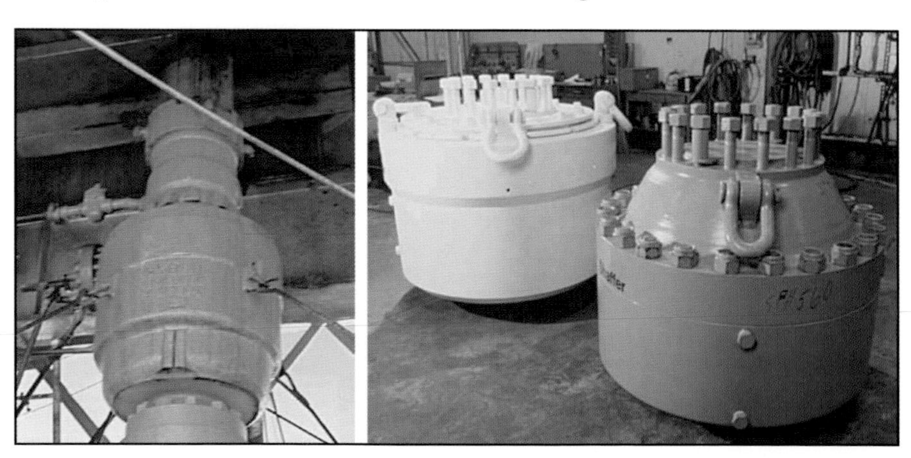

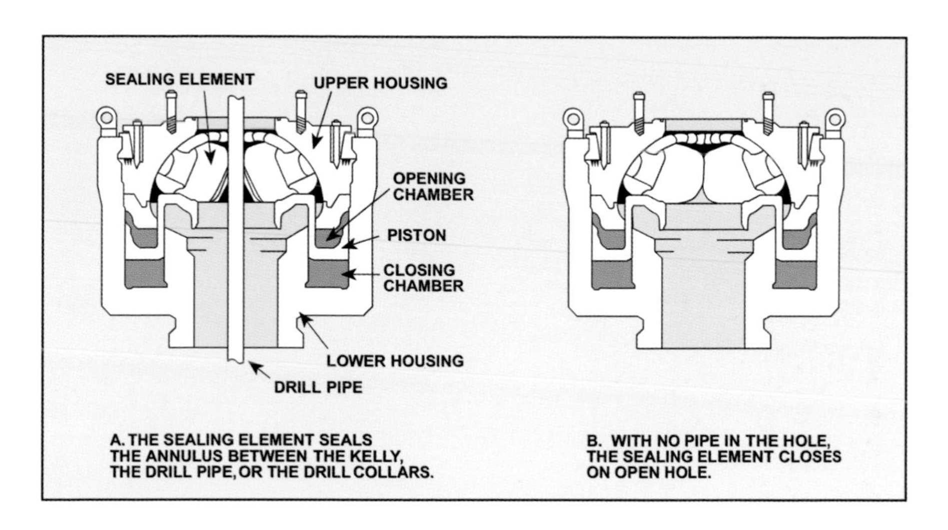

Annular Preventer

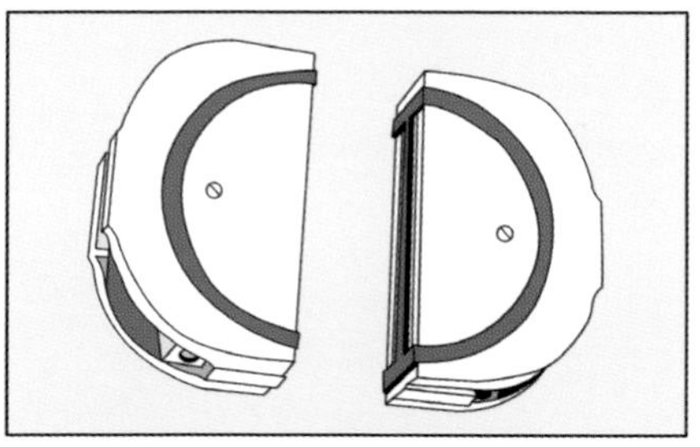

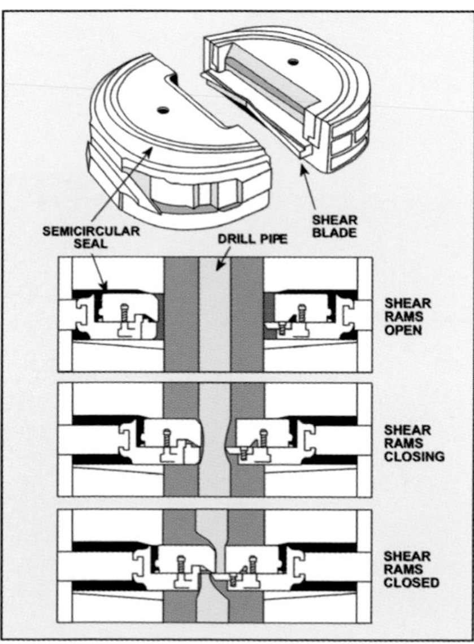

Blind shear rams

Blind Rams:
Blind rams are two large blocks of steel that provide a seal when there is no pipe in the well. This can be done because the surface of the blind rams is completely flat.

Pipe Rams and Variable-Bore Rams:
Pipe rams are similar to blind rams but fit only a particular set of pipes. As such, it is important for personnel to know what size of pipe the pipe rams can accommodate. Variable-bore rams can fit over a range of pipe and also provide a seal to the well.

Shear Rams:
Shear rams are designed to cut across the pipe and also provide a seal to the well. Shear rams are also designed to cut only a certain size of pipe, and specifications must also be known regarding the biggest size of pipe the shear rams can accommodate.

Accumulator:
Accumulators are a set of bottles that store pressure and allow the various rams to be activated. The bottles typically contain hydraulic fluid and nitrogen gas and are kept at a certain pressure with the use of pumps. In case the pumps fail, the preset pressure in the accumulators allows the activation and use of all rams. Maintaining a set amount of pressure in the accumulators is very important.

Choke/Kill Manifold:
Once the well is sealed either with the use of the annular preventers or any other rams, the well is then controlled with the use of the choke/kill manifold. Through a series of valves, mud is pumped down the well through the kill line, and the mud is returned to the surface through the choke line. The choke line contains a "choke" valve that controls the flow of mud, hence controlling the well through this choke valve. The choke/kill manifold is controlled through a control panel on the rig floor adjacent to the driller's console.

BOP Panel:
The BOP panel allows the driller, subsea engineer, or trained person to control the BOP from the rig floor. In addition, the status of each valve is also shown on the BOP panel.

4.1.2. Subsea BOP

The actual subsea BOP has the same components as a surface BOP, with quite a few more pieces of equipment for redundancy and control. What makes up the entire subsea equipment from the well up to the surface is listed; each part will be discussed in different sections of this chapter:

- BOP and LMRP
- riser
- telescopic joint
- tensioners
- diverter
- surface choke/kill manifold
- blue and yellow MUX reels
- hotline reel
- BOP control panel
- UPS
- accumulator room
- HPU room

Subsea BOP

Due to the distance between the BOP deployed on the seabed and the vessel being from 1,000 ft. to as high as 15,000 ft., subsea BOPs are operated through electrohydraulic components (a combination of electrical and hydraulic components). On a surface BOP, the short distance between the hydraulic components on the surface and the actual BOP allows for easy activation and response time. If the same method was used in subsea BOPs (only hydraulics), the response time would make it very inefficient for controlling the BOP's prime movers. As such, electrical signals are used alongside hydraulics. When a function is triggered from the BOP panel (opening or closing of a valve), the electrical signals travel to the BOP, and, using the accumulator bottles on the BOP, valves are opened or closed. This is the primary difference between a surface and subsea BOP. The added complexity is because of the need to have redundant systems.

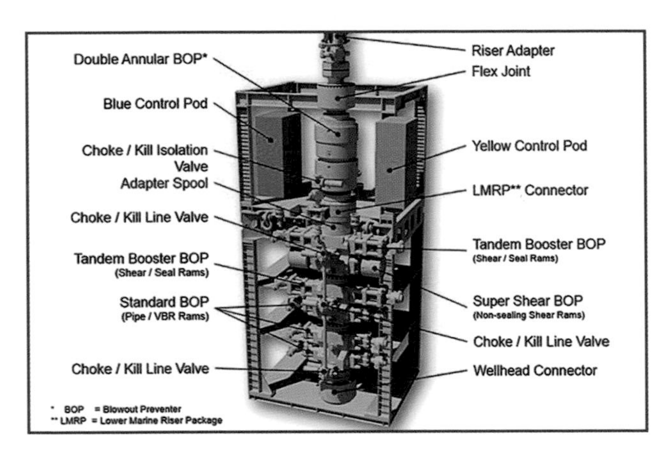

Subsea BOP with parts

Riser Adapter:

This provides the connection between the riser string and the BOP/LMRP, which eventually connects to the vessel on the surface. The adapter also has flexible hoses that connect to the choke/kill line and the conduit line. The conduit line is the means of filling up the accumulator bottles on the BOP with hydraulic from the surface.

Flex Joint:

Just as the name implies, the flex joint allows for movement from the vertical plane of the BOP without causing any damage to the BOP and the riser. Most flex joints are designed to allow movements by as much as 10° from the vertical relative to the BOP. Vessels are designed to disconnect the BOP from the LMRP well before this 10° angle is reached. As such, a vessel drilling in 5,000 ft. of water will have disconnected before reaching an excursion of approximately 880 ft. A lot of other factors come into play determining the disconnect point of the BOP from the LMRP, but the BOP's flex joint angle limitation is one of the determining factors.

Flex joint

Double-Annular BOP:
The annular here is the same as discussed in surface BOPs. The annular preventer, when closed, provides a seal around the pipe that is currently in the well, thereby "shutting in the well." Double annular implies that this BOP has an upper and a lower annular preventer. This is usually the first annular to be closed during a well control situation.

Blue/Yellow Control Pod:
The pods on the BOP contain two sections: the MUX section, which houses the electronic and pilot valves components; and the MUD section, which houses the hydraulic component. Within each of the blue and yellow pods are two independent subsea electronic modules (SEMs), named SEM A and SEM B. When the BOP is deployed and in use, one pod is active and in use while the other pod is inactive, and within the active pod, one SEM is in use. This is done for redundancy reasons. When a signal is sent by the operator on the vessel to the BOP (via MUX lines), both pods and all SEMs receive the signal, but only the active pod and SEM will fire its hydraulics, thereby activating a certain function. All pods and SEMs will therefore know what last function was activated, so that when the operator decides to switch from one pod to the other or one SEM to the other, the same last known function is carried throughout the system.

LMRP Connector:
The LMRP connector provides the connection between the BOP and the LMRP.

Choke/Kill Line Valve:
The kill line allows for mud to be pumped down the well when the well is shut in, and the choke line is used to circulate mud back up to the surface with the use of a throttle valve to regulate the pressure.

Shear / Variable Bore / Pipe Rams:
These have all been covered in the "Surface BOP" section. The difference in subsea BOPs is the number of RAMs that a subsea BOP will have. On the modern deepwater drillship there are typically seven RAM preventers.

Wellhead Connector:
The BOP finally connects to the wellhead through the wellhead connector. The wellhead connector contains a rubber seal set to a particular pressure to maintain a seal and connection to the wellhead.

4.2. Risers

Riser stored on deck

Once the well has been drilled to a certain depth, the BOP is lowered to the seafloor and connected to the wellhead.The riser provides the medium for drilling fluids and the drill string between the vessel on the surface and the existing well. Risers contain floats that can reduce the effective weight of the riser by 90%–95% once in the water, thereby reducing the weight that needs to be supported by the vessel. Risers are connected to each other by means of riser bolts, and between each riser there is a seal sub that provides a seal for fluids between two joints of riser. Risers vary in length but typically measure 50, 65, 70, 75, or 90 ft. Smaller joints of risers (called pup joints) are also used to properly adjust for the riser space out, depending on the water depth. Each joint of the riser contains a choke/kill line, mud booster line, and rigid conduit line. In addition to these, clamps are secured to the side of the riser and hold in place the MUX lines connected to the BOP from the surface.

To prevent a certain pressure differential between the seawater and the pressure inside the riser, certain riser systems contain a riser fill-up valve.This valve will automatically open and fill up the riser with seawater at a preset pressure, to prevent the riser from being crushed due to this preset differential pressure.

Various flow lines on riser visible

Riser being picked up

4.3. Telescopic Joint

The topmost riser joint that connects the vessel to the marine riser system is called the telescopic joint, which consists of an inner barrel and an outer barrel. A cross-sectional view of the telescopic joint will reveal these features (outer barrel and inner barrel); hence the term "telescopic." The top section of the telescopic joint will also have a flex joint. Similar to the flex joint above the BOP (called the lower flex joint), the flex joint above the telescopic joint (upper flex joint) provides lateral movement of the vessel without damaging the riser joint. Goosenecks are also connected to the telescopic joint, and from these, flexible hoses for the choke/kill line, rigid conduit lines, and booster lines connect the riser system to the systems on the vessel.

The inner barrel is attached to the rig floor and is rigidly connected to the vessel, while the outer barrel is connected to the entire riser string (all in the water) and is secured to the vessel by means of wires (called tensioners; see p.90). These wires can be adjusted by the vessel's crew to carry a certain amount of weight, and, depending on operations, either more or less weight is supported by the tensioners.

There are two seals (called packers) between the inner barrel and the outer barrel. The top seal is an air-pressurized packer, with the air coming from the vessel's air system (typically at 10–15 bars), while the lower packer is a hydraulically operated seal. Again, two means for redundancy and also one system that can provide a tighter seal over the other.

Slip joint

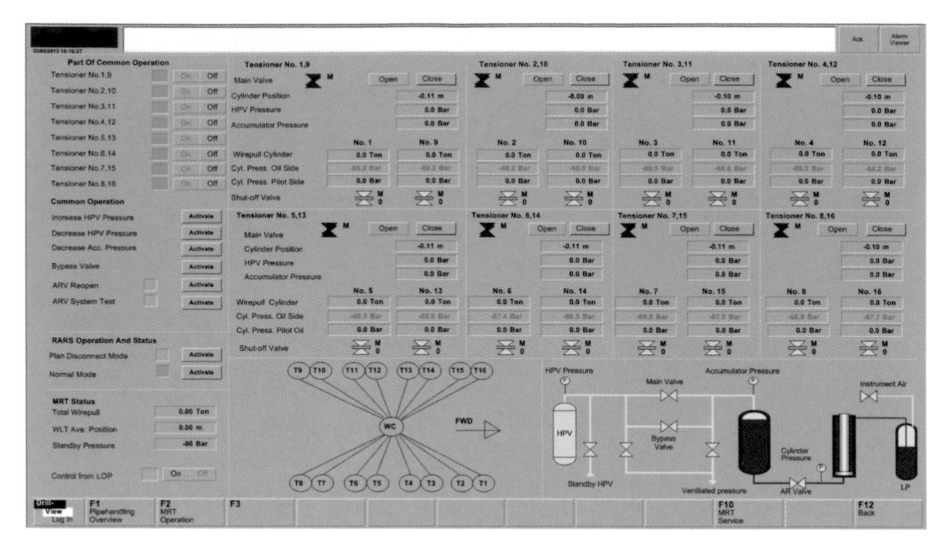

Tensioner control system

4.4. K.T. Ring / Fluid Bearing

The tension ring (known as a KT ring or fluid bearing by different manufacturers) allows the vessel to change heading (turning the vessel), depending on prevailing weather conditions, to provide for a better vessel ride. This is partly made possible by the tension ring, which rotates, thereby allowing the outer barrel to stay on the same heading while the vessel turns. Modern drillships can turn by as much as 180° on either side but will be limited because the hoses that are connected at the goosenecks start to pick up too much tension.

4.5. Tensioner System

Tensioners are wire ropes that support the weight of the riser system alongside all components beneath the water, through an attachment on the outer barrel of the slip joint. The weight of the joints at the top of the riser string can crush the joints below it; to prevent this buckling, a tensioner system is used in water depths greater than 300 ft. Crew members are capable of adjusting how much tension is held by the tensioner system, depending on variables such as

- water depth
- mud weight
- wave and current forces
- buoyancy effect of the riser
- vessel motion
- offset from well center

A tensioner system is made up of

- a tensioner ring attached to the outer barrel
- wire lines (typically 16)
- high-pressure accumulator
- sheaves that provide a path for the tensioner wire
- hydropneumatic cylinder with its sheave assembly
- control Panel

A single wire rope will be connected to the tension ring and will go through a set of sheaves from the moon pool, after which it is secured on the piston end of the tensioner unit. To maintain the preset tension placed by crew members as the vessel heaves, the piston within the tensioner unit will move, which corresponds to movement on the wire and hence the outer barrel of the slip joint. Knowing how much reeving your wire lines go through is important, since a corresponding movement on the piston within the tensioner unit will correspond to a certain movement of the actual wire line and hence the slip joint. The rod end of the cylinder above the piston is filled with fluid under a certain amount of pressure (20–40 psi), which helps lubricate the cylinder, dampens piston motion, and provides speed control in the event a wire is broken or the LMRP is disconnected from the BOP.

Left: Tensioner in moonpool

Right: Tensioner dead end

Deepwater drillship work limitations are also based on how much movement is allowed on the slip joint, and hence the tensioner unit. Drillships with a 30 ft. stroke limitation means that the vessel is allowed a 15 ft. movement on either side of the midpoint prior to stopping operations. If the heave at any moment of the slip joint exceeds 15 ft. in one direction, operations will have to stop and the LMRP disconnected from the BOP.

4.6. Choke and Kill Lines and Manifold

During a well control situation, the well is shut in and the choke and kill lines are used to control the flow of mud. When the well is said to be shut, it means the annular(s) are closed and the well is essentially shut; depending on the situation, this shut-in could either be with or without drill pipe in the hole. Pumping into the well while it is shut in is still achieved through the drill pipe but can also be done through the choke/kill line. On the surface, valves get lined up to pump down the kill line from the surface, through the riser, and into the BOP. The returns are then taken up the choke line on the BOP and back up the same line on the riser until the returns get to the surface.

Choke

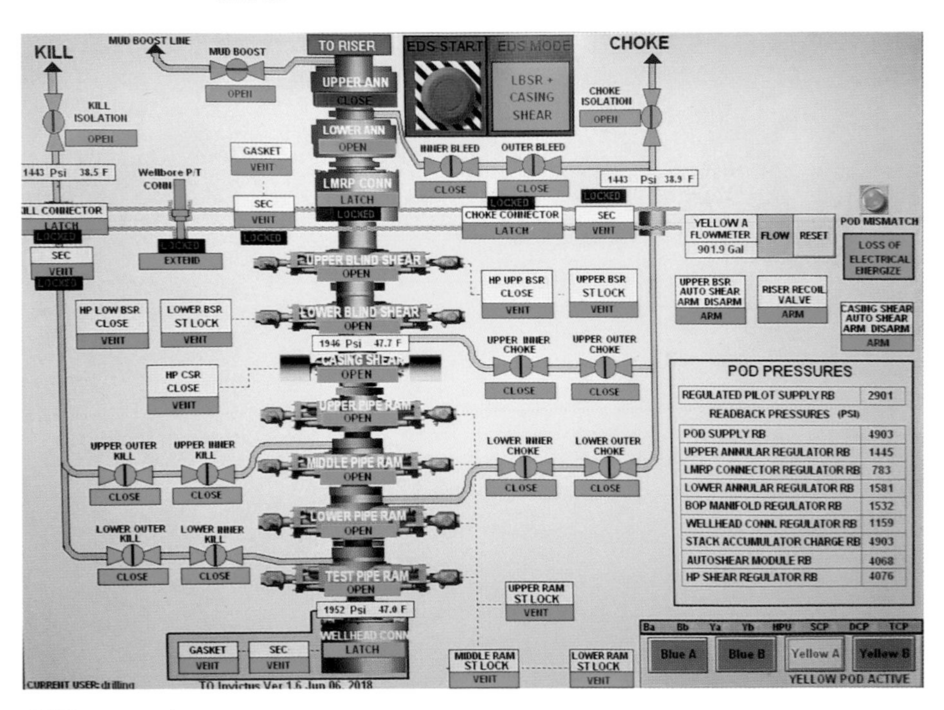

BOP operating system

Once on the surface, a choke is used to control the flow and pressure necessary to keep the well stable. This choke is controlled remotely by a choke panel on the rig floor.

4.7. UPS

The UPS (uninterrupted power supply) used for the BOP systems is very similar to UPS used in DP and drilling equipment. UPSs have the capacity to provide certain equipment with power from a battery bank in case there is loss of main vessel power. The two UPSs aboard modern drillships for BOP systems typically have a minimum capacity of two hours each and serve certain equipment, such as

- BOP control panels (all three)
- diverter control unit
- fluid-mixing units (for mixing your BOP fluids)
- hydraulic power unit
- event loggers
- blue and yellow control pods
- blue and yellow cable reels

Given that the BOP system is electrohydraulic, the main function of the BOP UPS is to keep the "brains" of the BOP alive during a full blackout—essentially your blue and yellow pods alongside the control panels. Given that these units still have power, any function carried out from the panel will still be executed, since your primary movers are controlled hydraulically. So, despite the BOP being a relatively big piece of equipment, multiple functions can still be achieved through the UPS even in the event of a blackout of the vessel.

4.8. Surface Hydraulics and Accumulators

Surface hydraulic equipment is primarily composed of mixing units, hydraulic power units (HPUs), and a diverter control unit. The mixing unit mixes up water, glycol, and a lubricant (BOP concentrates, also known as BOP fluid / stack magic), which is sent to the HPU and stored in a fluid reservoir tank. From the reservoir tank, hydraulics are sent both to the diverter control unit and the accumulators.

The diverter control unit is used to control functions of the diverter, send pressure to adjust the pressure on the packer between the inner and outer barrel (discussed in riser), and also send hydraulic fluid subsea if needed.

The accumulators are bottles filled both by the hydraulic fluid and an inert gas (typically nitrogen or helium) set at a certain pressure (around 4,800 psi); a bladder inside the bottle provides the medium within which

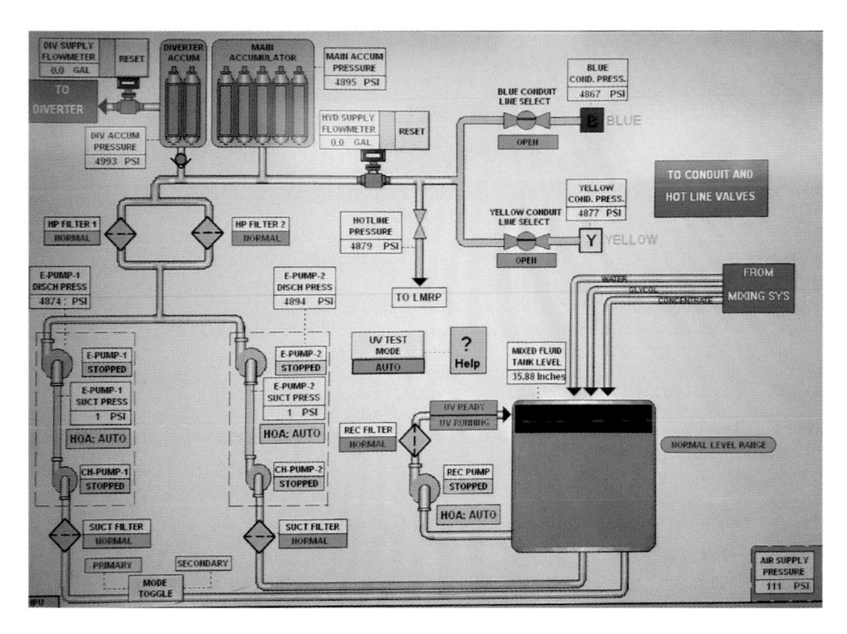

the gas pressurizes the fluid. Surface accumulators are primarily composed of a diverter accumulator (used for diverter functions) and the main accumulator.The main accumulator is in turn used to send fluid to the BOP, which is subsea, either through the rigid conduit line or blue or yellow conduit lines.The diagram depicts what an overview of the surface hydraulic equipment looks like.

4.9. BOP Panel

The BOP panel provides the interface between the equipment and the operator. Functions of surface and subsea components can be controlled from the BOP panel.

Certain surface equipment controlled from the BOP Panel includes but is not limited to

- diverter
- mixing pumps for BOP fluids
- pumps to pressure up surface and subsea accumulator bottles
- functioning valves for mixing BOP fluids
- functioning valves on the BOP

The major subsea components that can be functioned from the BOP panel are easily described by the picture depicting both the BOP and LMRP. Every valve and RAM can be functioned from the BOP panel and also initiate an emergency disconnect sequence (EDS). An EDS is used in cases of emergency whereby the BOP is disconnected from the LMRP and the well is sealed.

KILL
TO RISER
EDS START
EDS MODE
CHOKE
POD PRESSURES
YELLOW POD ACTIVE
CAMERON
CAMERON

BOP panel

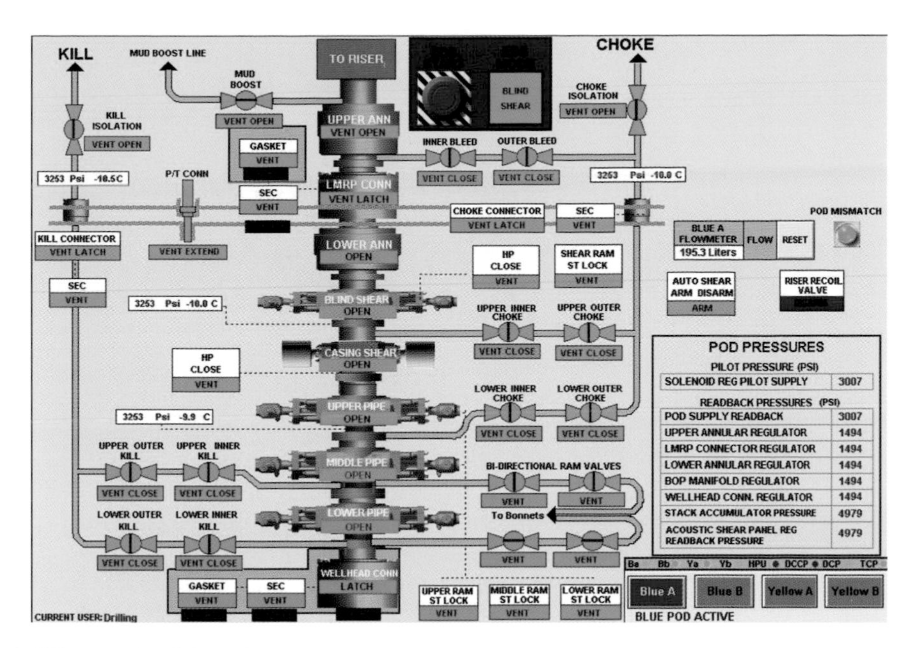

BOP panel with valves

For redundancy, each BOP panel has two screens, and modern drillships will contain at least three panels, typically in the bridge, on the rig floor, and in the subsea office (see p. 98). Every BOP panel will mimic the latest function and the status of each equipment. Key items from the BOP panel show the operator which pod is selected as the active pod (i.e., either blue pod or yellow pod); further, within each pod which subsea electronic module (SEM) is the active one, either A or B.

From the status indicator of the BOP panel, we know that the yellow pod is currently selected as active, and within that pod, SEM A is in control. The green light indicators further tell that each pod and SEM is functioning, and so are each of the BOP panels (DCCP, DCP, and TCP). Finally, the HPU unit is also in good condition. This indicator light is typically on the lower right corner of the BOP panel.

To carry out a required function, the operator needs to press down on the enable button in addition to selecting the desired function; again, this is done to prevent the inadvertent functioning of a piece of equipment.

4.10. Crown-Mounted Compensator (CMC)

The crown-mounted compensator (CMC) is a type of active heave compensator (AHC) mounted at the top of the derrick (hence the term "crown mounted"). Its main task is to compensate for the effects that heave (up-and-down movement of the entire vessel) has on the drill string during drilling operations. When the entire vessel heaves (from the effects of

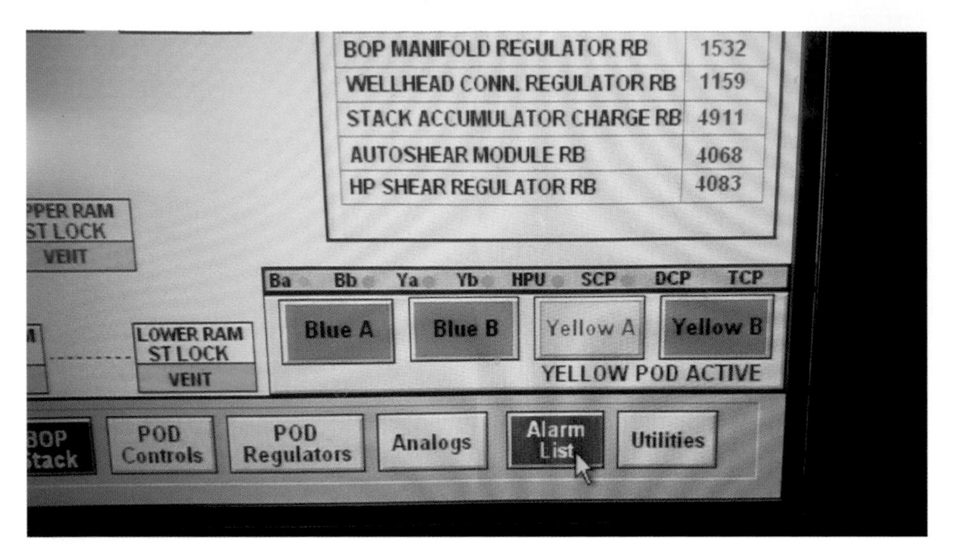

Active POD

Crown mounted compensator

weather) and the CMC is in use, an opposite movement from the CMC keeps the drill string in the same position. By doing so, the bit is spared from constant pitting on the bottom of the well, which damages the bit. A general picture of a CMC is depicted.

CMCs are primarily limited by how much compensating stroke (total height after which the CMC can no longer be used) and the weight that the entire CMC can support, called the dynamic compensating load, which is on the crown (refer to section 3.4, "Top Drive and Hoisting Equipment," as a refresher on crown). Modern drillships have a maximum compensating stroke of 25 ft. and a dynamic load of 700 megatons. This means that after a total heave of 25 ft., the CMC cannot be used, and if the weight on the drill string is more than 700 megatons, it also cannot be used. Though this is not typical subsea equipment, the equipment is maintained by the subsea engineer.

5 SUPPORTING OPERATION

The following operations will be found on any drillship, with the exclusion of some. Where certain operations are typically not found on the drillship, it will be pointed out as an exception, or possibly a rarity.

5.1. Cranes

Cranes offshore are primarily used for moving cargo from supply boats (which deliver goods to the offshore vessel) to the vessel and for moving equipment around the vessel. Typical deepwater drillships have five major cranes: one on each corner of the vessel and the fifth one on the stern. Certain drillships have gantry cranes designed to move risers on deck during riser operations. Each of these cranes is rated to carry about 100 tons and will typically include one crane capable of conducting subsea lift. The subsea crane is specialized and will have an active heave compensator (similar technology as the CMC on the rig floor); it can be rated for as much as 165 tons. If the vessel is rated to work in a particular water depth (12,000 ft, for modern drillships), the subsea crane will have the same capacity of 12,000 ft. water depth for subsea lifts. Though cranes are primarily used for moving equipment and conducting subsea lifts, they have also been used to deploy equipment over the side of the vessel, such as seismic-survey equipment. During certain exploratory phases the operator can decide to acquire more data from the well, using both drilling equipment and seismic guns deployed over the side of the vessel with the use of cranes.

Other cranes are used specifically for moving the BOP/LMRP from the moon pool to the BOP setback area—an area adjacent to the subsea engineer office to work on the BOP/LMRP. All of these cranes are electro-hydraulically operated.

Crane moving equipment on deck

BOP crane

The most versatile set of lifting equipment found around miscellaneous areas of the vessel is air-operated winches (termed tuggers). Tuggers will be on the rig floor, in the moon pool area, and other conspicuous areas around the vessel. These will be used by the crew to move equipment around their work area and can have a carrying capacity as high as 10 tons.

Tugger picture

5.2. ROV

A remotely operated vehicle (ROV) is essentially the eyes of the operations that take place subsea. ROV operators are typically not made up of crew members employed by the owner and operator of the drillship, but are contracted by the client to be on board. Drillships are typically outfitted with one ROV, but certain clients request more, and it would not be uncommon to find two ROVs on board. They are linked to the vessel by an umbilical cord and are deployed to the seabed while in a cage. Once the ROV operator has lowered the ROV cage to a set depth, the operator hovers the ROV to any desired position. ROVs are limited by their operating depth and the umbilical linking the ROV to its cage. ROVs are deployed with multiple equipment, such as sonars, cameras, manipulator arms, high-powered lights, and cutting arms, and are even capable of taking samples of water or particular fluids. In deepwater drilling operations they play critical roles, such as

- surveying the seabed before any drilling activity begins
- during drilling operations and before the BOP gets deployed, monitoring the seabed for any shallow gas
- deploying acoustic systems used as reference systems by dynamic-position operating systems
- when the BOP is deployed, inspecting the condition of the riser and assisting in troubleshooting the BOP by monitoring for leaks

Their manipulator arms have the capability of controlling certain BOP functions from the seabed, such as opening and closing certain rams, and even emergency disconnection of the BOP from the LMRP.

ROV being deployed

ROV on deck

5.3. Mud-Logging Unit

Similar to the ROV crew, mud loggers are typically not core members of the drillship but play an equally important role during drilling operations. Companies who run mud-logging units will install sensors throughout the drilling package (level sensors for various mud tanks, indications of mud pumps running, gas levels in various spaces and tanks, etc.) to capture everything that can affect drilling operations. All these data are analyzed and displayed in real time for personnel involved in drilling operations. By doing so, mud loggers actively track the status of the well and are another set of eyes during drilling operations. A key aspect of what they track is the pit volume totalizer (PVT). Simply tracking PVT can easily tell personnel the condition of the well. Mud loggers typically have certain criteria that, when met, require notification of the driller, tool pusher, drilling superintendent, and company representative. Mud loggers, similar to DPOs and drillers, stay in a control room that is always manned during drilling operations. Additional information displayed by the mud logging units includes

- weight that is being supported by the top drive
- length of the drill pipe in the hole
- status of each mud pump

- level in each mud tank
- mud weight going in and out of the well
- flow rate of mud coming out of the well
- gas levels of mud coming out of the well
- total pressure on the standpipe
- total pressure from the cement unit

This is not an exhaustive list of data displayed by mud loggers; when charted over time, they help personnel interpret what is going on with the well. The picture is an example of a display of a mud-logging unit.

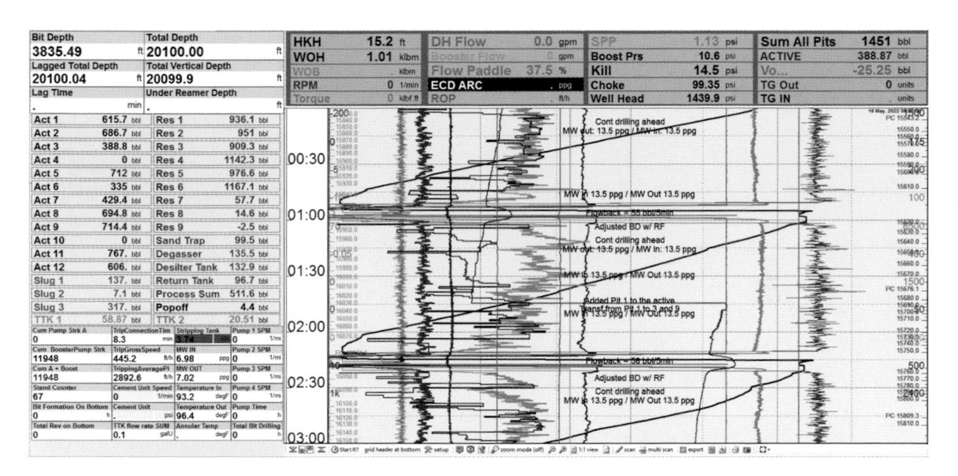

Mud logging data

5.4. Flaring/Discharge of Oil

Depending on what stage of the well the drillship is involved in, gas might be flared over the side of the ship through a flare boom, or certain piping on the vessel will discharge oil coming from the well into cargo tanks on the vessel or discharged to a vessel alongside the drillship.

People in the industry can go for years without being involved in such operations; it all comes down to whether the vessel has the equipment to complete these tasks. Most drillships today will have either flare booms or cargo tanks, or a combination of both. During flaring operations, gas is brought up to the surface and flared for a period of time by the operator. By doing so, operators can better estimate what the gas reserves are, and similar steps are done when oil is brought to the surface. Oil is not flared but instead is stored in storage tanks on the vessel and eventually discharged to another vessel.

Flare boom

5.5. Leisure

Even though-leisure activities do not comprise supporting operations, putting personnel at ease and giving them a sense of being at home is key in maintaining morale offshore. Certain facilities you will find in modern day drillships include

- gymnasium for the crew (some vessels have two due to their high capacity)
- TV rooms for the crew
- cinemas
- videos games in each individual's room
- external drive accessible from rooms for a variety of TV programs
- satellite TV in each room
- saunas

5.6. Intervention and Workover Control Systems (IWOCS)

During certain phases of developing a well, an operator might need to install certain equipment to function subsea equipment, or to complete intervention work, tree installation on a well, or even a nearby well while drilling operations are being conducted on a different well. A lot of coordination is needed from multiple parties to accomplish this, but it streamlines a lot of processes and allows clients to use existing deck space on a rig already on charter. Personnel might go through their careers without having to experience this, but it is not uncommon to have one on board; again, it depends on what stage of the well the client is in. These IWOCS systems are typically mounted on a lowering and retrieving arm (LARS) and have high-pressure hoses that run from the rig to the subsea manifold and in turn are used to complete certain functions on a wellhead or subsea tree. An example of a LARS unit deployed is depicted.

To the right side of the picture, you can see the LARS unit with the umbilical cable, and to the left you can see the umbilical spool on deck.

LARS unit on deck

5.7. Wire-Line Units

Once the well has been drilled to a certain depth, a wire line is used to lower equipment into the well, and data about the well are recorded at predetermined depths. Wire line can also be used to retrieve tools lost in the well, or for well intervention. The wire-line unit consists of a container typically aft or adjacent to the aft conveyor (aft of the rig floor). Adjacent to the container will be a drum that has a very fine wire line (hence the term "wire-line unit") that varies between 2 and 4 mm in diameter. The wire line runs from this drum to a sheave on the rig floor and from the sheave down into the well.

Wire-line unit

5.8. Current Meter

Current meter

Another piece of equipment found on some rigs is a current meter. This typically consists of one unit mounted on a frame on the vessel and lowered into the water to a depth of about 30 ft., and another unit lowered to the seabed by the ROV. Combined, these two units calculate what the current is at different water depths, which is displayed on a screen at the surface. Personnel are able to view a profile of the current. This becomes very useful when drilling in areas with high current, or when the rig encounters loop currents. In high current, the rig can be deliberately offset to compensate for the effect of current on the riser and the drill pipe. This is all done to reduce the effective angle of the riser in relation to the BOP.

5.9. Weather Radar

Similar to radars used on land to assist in determining the weather approaching, rigs are equipped with weather radars for the same purpose. Weather radars are capable of estimating the wind forces that could affect the rig, the intensity of rain, the speed at which a front is moving, and the approximate size of incoming weather. All these data help the DPO on the bridge to determine an ideal heading and steps to take to get the vessel ready for incoming weather. With an ideal heading selected, modern drillships might encounter an offset (being moved) of no more than 10 ft. if wind changes from 10 knots to 50+ knots; again, this depends on whether the operator has taken all precautionary measures.

5.10. Offshore Supply Vessels

Though offshore supply vessels (OSVs) are not in any way part of the drillships, it is important to note that it is through OSVs that the drillship gets all its supplies, from fuel to groceries to material needed to drill a well. OSVs are contracted by the same company contracting the drillship, and a set OSV will make runs between the shore and the location of the drillship. The picture depicts ongoing cargo operations between an OSV and a drillship.

OSV alongside for cargo ops

6 TECHNOLOGICAL ADVANCES/NEW DEVELOPMENTS

6.1. Storage Units below Deck

Though storage of equipment or cargo below deck is the norm for regular cargo-carrying vessels, this is a little uncommon aboard deepwater drilling units. It is particularly hard to achieve for stability reasons and the general layout of the vessel. Over the last few years, more vessels have been designed to stow the riser below deck. Risers are used only when lowering and retrieving the BOP, and storing it below deck creates room and deck space to be used to store other equipment for the well operator. Doing so reduces trips made by supply vessels and in the long run can be a cost savings to the vessel contractor if more equipment needed for the well is stowed on deck. Furthermore, storing the riser below deck allows the subsea engineer to have easier access to each joint of riser for maintenance and inspection.

It is the author's point of view that deck stowage is an area where there could be improvements in the near future, either with drillships having a bigger variable deck load (how much can be loaded onto the vessel) or better layout for the stowage of tools and equipment used offshore.

Riser stored below deck

6.2. Hook Load

A key limiting factor for deepwater well operations is the hook load; this is how much weight can be supported by the top drive during any operations.The heaviest hook load on board deepwater drillships is encountered when running casing of a certain size. Seventh-generation drill ships can support up to 2.5 million lbs.; this limitation has been pushed to 3.0 million lbs. recently. Heavier hook load provides more options for well operators, so this is a frontier that will constantly be pushed. Heavier hook loads means that a well could potentially be drilled a little deeper prior to running a certain casing size, and by doing so, a well could potentially be drilled faster if fewer trips are made for running casing or trips for drilling.

6.3. Managed Pressure Drilling

Managed pressure drilling (MPD) has been around for several years but is becoming more prevalent on board deepwater drillships. In a nutshell, MPD allows a constant pressure to be maintained at the bottom of the well during all drilling operations. On conventional vessels, when a drill pipe is being connected on the rig floor, mud pumps have to be shut off to connect another joint of pipe with the pipe already in the hole. By doing so, the bottom of the well encounters a momentary loss of pressure, primarily due to the mud pumps being shut off.This could be very delicate, especially when drilling wells with minimal pressure tolerances. MPD has allowed many operators to access some wells that at first were hard or impossible to access due to these tight pressure tolerances.The exact functionality of MPD goes beyond the scope of this book, but multiple modifications are done both on the vessel and when deploying a special joint of riser called the MPD joint.

6.4. Draw Works Compensated

The crown-mounted compensator (CMC) (see 4.11) requires a lot of equipment to be above the rig floor (cranes, sheaves, pistons, and a multitude of equipment), not only making the vessel "top heavy," but also increasing the risk of equipment getting damaged and falling to the rig floor. Personnel need to go to the top of the derrick regularly for maintenance, further increasing the risk not only to personnel working above but personnel below. As a refresher, the CMC compensates for the heave (up-and-down movement) of the vessel while drilling, thereby keeping the drill bit at the required depth. This is now being achieved on certain vessels by having the compensation occurring on the draw works; as the vessel moves up, more wire is paid out, and wire is taken in when the vessel moves down.

6.5. Electrical BOP

There have been a lot of changes regarding BOP, the most notable being having a fully electric BOP rather than a hydroelectric BOP. This system reduces complexity and moving parts, thereby increasing reliability for well operators and the owners of the vessel. Also, BOPs discharge a lot of BOP fluid whenever they are functioned while at sea, and a fully electric BOP will do away with BOP fluid being discharged; a set of accumulators found in typical BOPs are replaced by a battery bank. Though these systems are still in their early stages, certain operators have designed and tested them; they have yet to be put to full use or trials.

6.6. Inertia Navigation

Inertia navigation systems have been used for several decades on airlines and in defense systems for missile guidance and are gradually finding their way into the offshore-drilling industry. Remember that the drillship stays on location through the use of dynamic-position systems that use a combination of satellite systems and acoustic systems on the seabed. Inertia navigation will enhance the capability of these systems, especially the acoustic system. While the DP system receives information on the location of the vessel every second from the satellites, depending on the water depth, the acoustic system can take as long as six to seven seconds to get an updated location; in critical situations these six to seven seconds can be valuable, and inertia navigation is being used to reduce this timing to match the GPS system.

6.7. Hybrid Power

A major expense for offshore drilling goes to paying for the fuel used on a daily basis to keep engines running. In an effort to reduce the carbon footprint that vessels emit, certain tugboats today are designed to run on battery during certain operations, and some are fully battery powered. Similarly, drillships are being redesigned to run generators more efficiently, with either a combination of battery and diesel generators or a lesser number of diesel generators running at a certain time. As technology evolves, this will be an area of improvement that drilling contractors will have to explore to reduce their carbon footprint.

6.8. Kinetic Blowout Stopper (K-BOS)

The kinetic blowout stopper, or K-BOS, is an electrically initiated, pyro-mechanical gate valve that performs the critical function of shearing and sealing a well in the event of a blowout. Certain subsea BOPs have been retrofitted with KBOS and tested to be able to shear certain pipe sizes in a much-faster time than the shear rams would have. This could be a potential game changer in the industry, since it provides an added reliability, and the quick time within which the shearing and sealing occur gives the operator more time to react to whatever issues they might be dealing with.

CONCLUSION

Deepwater drilling has been around for multiple decades and will continue to be as the world seeks to recover oil and gas in various regions of the world. Technological advances such as higher BOP ratings (20,000 psi), heavier hook loads, and MPD will make previously inaccessible basins accessible. As the industry continues to go through a boom-and-bust cycle, personnel will depart the industry and never return. For those venturing into this exciting industry, I hope this book has done a good job at painting a picture of the critical areas that make these drilling vessels capable of safely exploring and recovering hydrocarbons from the seabed.

GLOSSARY

abaft (*adv.*): 1. Toward the stern of a ship or mobile offshore drilling rig. 2. Behind. 3. Farther aft than.

abeam (*adv.*): To or at the side of a ship, vessel, or offshore drilling rig, especially at right angles to the ship, vessel, or rig's length.

accumulator (*n.*): 1. A vessel or tank that receives and temporarily stores a liquid used in a continuous process in a gas plant. 2. On a drilling rig, an assembly of devices such as bottles, control valves, pumps, and hydraulic fluid reservoirs that store hydraulic fluid under pressure and provide a way for personnel to operate (open and close) the blowout preventers. Also called surge bottles.

accumulator bottle (*n.*): A bottle-shaped steel cylinder in a blowout preventer control unit to store nitrogen and hydraulic fluid under pressure (usually at 3,000 psi). The fluid operates the blowout preventers.

acoustic backup system (*n.*): In offshore drilling from a floating drilling rig using a subsea blowout preventer stack, devices that send acoustic signals to a subsea receiver to operate the blowout preventer (BOP) components. The system consists of a miniature control pod with several subsea pilot-manipulated (SPM) valves to operate the selected BOP components. The system is used when the hydraulically operated system fails.

acoustic position reference (*n.*): A system consisting of one or multiple beacons positioned on the seafloor to transmit an acoustic signal to a hydrophone (or multiple hydrophones) mounted on the hull of a floating offshore drilling vessel to receive the signal, and a position display unit to track the relative positions of the rig and the drill site.

active heave compensator (*n.*): A hydraulic power-assist device used to overcome friction of the passive heave compensator seal and the drill string, which act in an opposite direction to the motion of the ship. Such compensators are used on board drillships and semisubmersible drilling rigs.

active mud tank (*n.*): One of usually two, three, or more mud tanks that hold drilling mud that is being circulated into a borehole during drilling. They are called active tanks because they hold mud that is currently being circulated.

adjustable choke (*n.*): A choke in which the position of a conical needle, sleeve, or plate may be changed with respect to their seat to vary the rate of flow; may be manual or automatic.

American Bureau of Shipping (ABS) (*n.*): A leading ship classification society, its main purpose is to determine the structural and mechanical fitness of ships and other marine structures through classification procedures. ABS establishes and administers standards, known as rules, for the design, construction, and operational maintenance of marine vessels and structures.

American Petroleum Institute (API) (*n.*): An oil trade organization (founded in 1920) that is the leading standardizing organization for oil field drilling and producing equipment. Headquartered in Washington, DC, it publishes materials concerning exploration and production, petroleum measurement, marine transportation, marketing, pipelining, refining, safety and fire protection, storage tanks, valves, training, health and environment, policy, and economic studies.

amidships (*n.*): The halfway point in the overall length of a floating offshore drilling rig or vessel.

anchor-handling tug / supply vessel (AHTS) (*n.*): A combined supply and anchor-handling ship. The AHTS is an offshore supply vessel specially designed to provide anchor-handling services and to tow offshore rigs, platforms, barges, and production modules / vessels. The vessel is often used as a standby rescue vessel for oil fields in production. The AHTS is often equipped for firefighting, rescue operations, and oil recovery.

annular (*adj.*): Pertaining to the annulus, which is sometimes referred to as the annular space.

annular blowout preventer (*n.*): A large valve, usually installed above the ram blowout preventers, that, when closed, forms a seal in the annular space between the pipe and the wellbore or, if no pipe is present, in the wellbore itself.

annular space (*n.*): 1. The space that surrounds a cylindrical object within a cylinder. 2. The space around a pipe in a wellbore, the outer wall of which may be the wall of either the borehole or the casing; sometimes termed the annulus.

appraisal well (*n.*): A well drilled to confirm and evaluate the presence of hydrocarbons in a reservoir that have been found by a wildcat well.

automatic choke (*n.*): An adjustable choke that is power-operated to control pressure of flow.

background gas (*n.*): In drilling operations, gas that returns to the surface with drilling mud in measurable quantities but does not cause a kick. Increases in background gas may indicate that the well is about to kick or has kicked.

ballast (*n.*): For ships, water taken on board into specific tanks to permit a proper angle of repose of the vessel in the water, and to ensure structural stability.

ballast tank (*n.*): Any shipboard tank or used for carrying saltwater ballast.

ball joint (*n.*): A device mounted between the annular preventer and the riser adapter on the lower marine riser package (LMRP) to prevent excessive bending forces from being exerted on the marine riser, the LMRP, and the blowout preventer components. It is a forged steel ball and socket containing a cylindrical neck extension with a riser adapter attached at the top of the neck.

bareboat charter (*n.*): An agreement to lease, or charter, a ship or offshore drilling unit without crew or equipment, usually for a predetermined time.

barite (*n.*): Barium sulfate, $BaSO_4$; a mineral frequently used to increase the weight or density of drilling mud. Its relative density is 4.2 (i.e., it is 4.2 times denser than water).

barite slurry (*n.*): A mixture of barium sulfate, chemicals, and water of a unit density between 18 and 22 lbs. per gallon.

barrier (*n.*): In well control and underbalanced drilling, a means of preventing the uncontrolled flow of fluids from a well. Barriers range from drilling mud whose hydrostatic pressure is greater than formation pressure, to mechanical devices such as blowout preventers and drill stem valves, which, when closed, prevent the exit of fluids from the well.

bit (*n*): The cutting or boring element used in drilling oil and gas wells. The bit consists of a cutting element and a circulating element. The cutting element comprises steel teeth, tungsten carbide buttons, industrial diamonds, or polycrystalline diamonds (PDCs). These teeth, buttons, or diamonds penetrate and gouge or scrape the formation to remove it. The circulating element permits the passage of drilling fluid and utilizes the hydraulic force of the fluid stream to improve drilling rates. In rotary drilling, several drill collars are joined to the bottom end of the drill pipe column, and the bit is attached to the end of the drill collars. Drill collars provide weight on the bit to keep it in firm contact with the bottom of the hole. Most bits used in rotary drilling are roller cone bits, but diamond bits are also used extensively.

bit sub (*n.*): A sub inserted between the drill collar and the bit. (*see* sub)

blind ram (*n.*): An integral part of a blowout preventer that serves as the closing element on an open hole. Its ends do not fit around the drill pipe but seal against each other and shut off the space below completely. *See* ram.

block (*n.*): Any assembly of pulleys on a common framework; in mechanics, one or more pulleys, or sheaves, mounted to rotate on a common axis. The crown block is an assembly of sheaves mounted on beams at the top of the derrick or mast. The drilling line is reeved over the sheaves of the crown block alternately with the sheaves of the traveling block, which is hoisted and lowered in the derrick or mast by the drilling line. When elevators are attached to a hook on a conventional traveling block, and when drill pipe is latched in the elevators, the pipe can be raised or lowered in the derrick or mast.

blowout (*n.*): An uncontrolled flow of gas, oil, or other well fluids into the atmosphere or into an underground formation. A blowout may occur when formation pressure exceeds the pressure applied to it by the column of drilling fluid, and rig crew members fail to take steps to contain the pressure. Before a well blows out it kicks; thus a kick precedes a blowout. *See* kick.

blowout preventer (BOP) (*n.*): One of several valves installed at the wellhead to stop (prevent) the escape of pressure either in the annular space between the casing and the drill pipe or in an open hole (i.e., hole with no drill pipe) during drilling or completion operations. Blowout preventers on land rigs are normally beneath the rig, at or slightly below the land's surface; on jack-up or platform rigs, at the water's surface; and on floating offshore rigs, on the seafloor. *See* annular blowout preventer, ram blowout preventer, wire-line preventer.

blowout preventer control panel (*n.*): Controls, usually near the driller's position on the rig floor, that are manipulated to open and close the blowout preventers

blowout preventer (BOP) stack (*n.*): An assembly of blowout preventers placed on top of each other (stacked one on top of the other), typically consisting of one or two annular preventers and three, four, or more ram preventers. *See* annular blowout preventer, blowout preventer, ram preventer.

bottomhole (*n.*): The lowest or deepest part of a well. (*adj.*): Pertaining to the bottom of the wellbore.

bottomhole assembly (BHA) (*n.*): The portion of the drilling assembly below the drill pipe. It can be very simple—composed of only the bit and drill collars—or it can be very complex and made up of several drilling tools.

bottom plug (*n.*): A cement wiper plug that precedes cement slurry down the casing. The plug wipes drilling mud off the walls of the casing and prevents it from contaminating the cement. *See* cementing, wiper plug.

bulkhead (*n.*): An interior wall that subdivides a ship or a mobile offshore drilling rig into compartments.

bulk tank (*n.*): On a drilling rig, a large metal bin that usually holds a large amount of a certain mud additive, such as bentonite, used in large quantities in the makeup of the drilling fluid. Also called a P-tanks.

buoyancy (*n.*): The apparent loss of weight of an object immersed in a fluid. If the object is floating, the immersed portion displaces a volume of fluid, the weight of which is equal to the weight of the object.

buoyant riser joint (*n.*): In drilling from offshore floating rigs in deep water, a marine riser joint to which has been added syntactic foam modules. The syntactic foam modules, which contain many hardened spheres of air that are embedded in a rugged, foamlike material, are attached to the riser joint. The spheres provide buoyancy. Buoyant riser joints are required when drilling in deep water (depths beyond 4,000 ft. [1,200 m]) to relieve stress on the riser-tensioning system.

cable (*n.*): 1. A rope of wire, hemp, or other strong fibers; often, but not always, personnel in the petroleum industry call it cable wire rope. 2. In electronics, a wire made of a conducting material such as copper or aluminum. Usually cable is braided from several single strands of wire into a single conductor that is easier to bend than a solid wire of the same gauge.

caliper logging tool (*n.*): A device lowered into a wellbore to measure and record the wellbore's diameter.

cap gas (*n.*): Natural gas trapped in the upper part of a reservoir and remaining separate from any crude oil, salt water, or other liquids in the well.

cascade system (*n.*): In systems supplying a breathable source of air to workers wearing breathing equipment in a toxic atmosphere, a serial connection of air cylinders in which the output of air from one adds to that of the next.

casing (*n.*): Steel pipe placed in an oil or gas well to prevent the wall of the hole from caving in, to prevent movement of fluids from one formation to another, and to improve the efficiency of extracting petroleum if the well is productive. Most casing joints are manufactured to specifications established by API, although non-API-specification casing is available for special situations. Casing manufactured to API specifications is available in three length ranges. A joint of range 1 casing is 16 to 25 ft. (4.8–7.6 m) long, a joint of range 2 casing is 25 to 34 ft. (7.6–10.3 m) long, and a joint of range 3 casing is 34 to 48 ft. (10.–14.6 m) long. The outside diameter of a joint of API casing ranges from 4 to 20 in. (114.3–508.0 mm). Casing is made of many types of steel alloy, varying in strength and corrosion resistance.

casing crew (*n.*): The employees of a company that specializes in preparing and running casing into a well. Usually the casing crew makes up the casing as it is lowered into the well; the regular drilling crew also assists the casing crew in its work.

casing hanger (*n.*): A circular device with a frictional gripping arrangement of slips and packing rings used to suspend casing from a casing head in a well.

cathodic protection (*n.*): A method of protecting a metal structure from corrosion by making its surfaces cathodic and controlling the location of anodic areas so that corrosion damage can be reduced to tolerable levels.

cement (*n.*): A powder consisting of alumina, silica, lime, and other substances that hardens when mixed with water. Extensively used in the oil industry to bond casing to the walls of the wellbore.

cement additive (*n.*): A material added to cement to change its properties. Chemical accelerators, chemical retarders, and weight reduction materials are common additives.

cementing (*n.*): The application of a liquid slurry of cement and water to various points inside or outside the casing.

cementing pump (*n.*): A high-pressure pump used to force cement down the casing and into the annular space between the casing and the wall of the borehole.

choke (chk) (*n.*): A device with an orifice installed in a line to restrict the flow of fluids. Surface chokes are part of the Christmas tree on a well and contain a choke nipple, or bean, with a small-diameter bore that serves to restrict the flow. Chokes are also used to control the rate of flow of the drilling mud out of the hole when the well is closed in with the blowout preventer and a kick is being circulated out of the hole.

choke-and-kill system (*n.*): An assembly of lines (pipes), valves, and operating devices (a system) on a subsea blowout preventer that provide a conduit for drilling mud to be pumped into the well through a kill line, and a conduit for fluids exiting from the well to flow through a choke line to the surface.

choke-and-kill valve (*n.*): A device that when opened allows drilling mud to be pumped into the well, and drilling and other fluids returning from the well to be directed to the choke manifold on the surface. *See* choke line, kill line.

choke line (*n.*): A pipe, or conduit, usually installed on outlets below the ram blowout preventers, through which fluids in the annulus can flow to the surface and through the choke manifold when valves to it are open. On floating offshore rigs, flow may be directed either down the annulus or up the annulus; on most land rigs flow is only upward.

choke manifold (*n.*): An arrangement of piping and special valves called chokes. In drilling, mud is circulated through a choke manifold when the blowout preventers are closed. In well testing, a choke manifold attached to the wellhead allows flow and pressure control for test components downstream.

coaxial cable (coax) (*n.*): Electrical conducting wire (cable) that consists of a center conducting wire covered by polyethylene insulation. A second conductor of braided copper (or other conductor such as aluminum) fits over the polyethylene insulation. Finally, a vinyl cover protects the braided conductor. The braided conductor protects, or shields, the center conductor from extraneous and undesirable electrical interference that may affect the signal being carried by the center conductor.

cofferdam (*n.*): The empty space between two bulkheads that separates two adjacent compartments. It is designed to isolate the two compartments from each other, thereby preventing the liquid contents of one compartment from entering the other in the event of the failure of the bulkhead of one to retain its tightness.

Coflexip hose (*n.*): A brand name of a flexible, high-strength, steel-braided, high-pressure hose that conducts fluids on a drilling rig and other petroleum installations. Many manufacturers make this type of hose. It is very strong and has high crush, abrasion, and fatigue resistance. It is used for choke and kill lines, cementing lines, surface blowout preventer control lines, injection lines, and other applications.

coiled tubing (*n.*): A continuous string of flexible steel tubing, often hundreds or thousands of feet long, that is wound onto a reel, often dozens of feet in diameter. The reel is an integral part of the coiled tubing unit, consisting of several devices that ensure the tubing can be safely and efficiently inserted into the well from the surface. Because tubing can be lowered into a well without having to make up joints of tubing, running coiled tubing into the well is faster and less expensive than running conventional tubing. Rapid advances in the use of coiled tubing make it a popular way to run tubing into and out of a well.

company man (*n.*): *See* company representative.

company representative (*n.*): An employee of an operating company who supervises the operations at a drilling site or well site and coordinates the hiring of logging, testing, service, and work-over companies. Also called a company man.

conduit (*n.*): 1. A pipe, line, or channel through which fluids flow. 2. Solid or flexible metal or other tubing through which insulated electric wires are run.

confined space (*n.*): A space large enough and so configured that an employee can bodily enter and perform work, but (1) has limited entry or exit, (2) has insufficient natural ventilation, (3) could contain hazardous materials, and (4) is not intended for continuous occupancy.

coring (*n.*): The process of cutting a vertical, cylindrical sample of the formations encountered as an oil well is drilled. The purpose of coring is to obtain rock samples, or cores, in such a manner that the rock retains the same properties that it had before it was removed from the formation.

crossover joint (*n.*): A length of casing with one thread on the field end and a different thread in the coupling, used to make a changeover from one thread to another in a string of casing.

crown block (*n.*): An assembly of sheaves mounted on beams at the top of the derrick or mast and over which the drilling line is reeved.

crown saver (*n.*): A device mounted near the draw-works drum to keep the driller from inadvertently raising the traveling block into the crown block. A probe senses when too much line has been pulled onto the drum, indicating that the traveling block is about to strike the crown. The probe activates a switch that simultaneously disconnects the draw works from its power source and engages the draw-works brake.

cuttings (*n., pl.*): The fragments of rock dislodged by the bit and brought to the surface in the drilling mud. They are not the same as cavings, which are particles that fall off the wall of the hole. Washed and dried cuttings samples are analyzed by geologists to obtain information about the formations drilled.

cuttings-sample log (*n.*): A record of hydrocarbon content in cuttings gathered at the shale shaker; usually recorded on the mud log.

dampener (*n.*): An air- or inert-gas-filled device that minimizes pressure surges in the output line of a mud pump. Sometimes called a surge dampener.

darcy (*pl.*, **darcys**) (*n.*): A unit of measure of permeability. A porous medium has a permeability of 1 darcy when differential pressure of 1 atmosphere across a sample 1 cm long and 1 cm^2 in cross section will force a liquid of 1 centipoise of viscosity through the sample at the rate of 1 cm^3 per second. The permeability of reservoir rocks is usually so low that it is measured in millidarcys.

day rate (*n.*): An hourly or daily contract price the operator agrees to pay for use of rig, crew, and specified equipment. A day rate contract allows the operator to directly supervise the daily drilling operations.

day tank (*n.*): A fuel tank in the fuel supply system for a diesel engine, between the main supply tank and the engine, that holds a limited amount of fuel.

deadline (*n.*): The drilling line from the crown block sheave to the anchor, so called because it does not move. *Compare* fast line.

deadweight ton (dwt) (*n.*): The total carrying capacity of a ship or offshore floating rig expressed in long tons (2,240 lbs.); the ship's or rig's deadweight is the displacement of the fully loaded ship or rig less the weight of the ship or rig itself.

deepwater drilling (*n.*): A relative term that refers to offshore drilling operations in deep oceans or seas—currently about 5,000 ft. (1,500 m) and deeper. It presents a number of special problems related to water depth.

degasser (*n.*): The device used to remove unwanted gas from a liquid, especially from drilling fluid.

derrick (*n.*): A large, load-bearing structure, usually of bolted construction. In drilling, the standard derrick has four legs standing at the corners of the substructure and reaching to the crown block. The substructure is an assembly of heavy beams used to elevate the derrick and provide space to install blowout preventers, casing heads, and so forth.

derrick hand (*n.*): The crew member who handles the upper end of the drill string as it is being hoisted out of or lowered into the hole. On a drilling rig, they are also responsible for the circulating machinery and the conditioning of the drilling or work-over fluid.

derrickman (*n.*): *See* derrick hand.

desander (*n.*): A centrifugal device for removing sand from drilling fluid to prevent abrasion of the pumps. It may be operated mechanically or by a fast-moving stream of fluid inside a special cone-shaped vessel, in which case it is sometimes called a hydrocyclone. *Compare* desilter.

desilter (*n.*): A centrifugal device for removing very fine particles, or silt, from drilling fluid to keep the amount of solids in the fluid at the lowest possible point. Usually the lower the solids content of mud, the faster the rate of penetration. The desilter works on the same principle as a desander. *Compare* desander.

development drilling (*n.*): Drilling that occurs after the initial discovery of hydrocarbons in a reservoir. Usually several wells are required to adequately develop a reservoir.

development well (*n.*): 1. A well drilled in proven territory in a field to complete a pattern of production. 2. An exploitation well.

directional driller (*n.*): An employee of a directional-drilling-service company whose main job is to help the driller at a well site keep the wellbore as close as possible to its planned course.

directional drilling (*n.*): Intentional deviation of a wellbore from the vertical. Although wellbores are normally drilled vertically, it is sometimes necessary or advantageous to drill at an angle from the vertical. Controlled directional drilling makes it possible to reach subsurface areas laterally remote from the point where the bit enters the earth. It often involves the use of deflection tools.

discovery well (*n.*): The first oil or gas well drilled in a new field that reveals the presence of a hydrocarbon-bearing reservoir. Subsequent wells are development wells.

diverter (*n.*): In offshore drilling, an assembly of devices used to direct fluids flowing from a well away from the drilling rig. When a kick is encountered at shallow depths, the well often cannot be shut in safely, because shutting it in on a shallow formation may create pressures high enough to fracture (break down) the formation, so a diverter is used. When activated, it allows well fluids to flow through a side outlet to a line (pipe) that carries the well fluids a safe distance away from the rig. A diverter contains a packing element, flow-line seals, and lockdown mechanisms and is run and retrieved with a special handling tool made up on riser pipe.

doghouse (*n.*): A small enclosure on the rig floor used as an office for the driller and his crew.

dogleg (*n.*): 1. An abrupt change in direction in the wellbore, frequently resulting in the formation of a key seat. 2. A sharp bend permanently put in an object such as a pipe, wire rope, or wire rope sling.

dope (*n.*): A lubricant for the threads of oil-field tubular goods. (*v.*): To apply thread lubricant.

downtime (*n.*): 1. Time during which rig operations are temporarily suspended because of repairs or maintenance. 2. Time during which a well is off production.

draft (*n.*): The vertical distance between the bottom of a vessel floating in water and the waterline.

draft mark (*n.*): Numbers that show the distance from the bottom of the keel or the lowest projection on a rig to the waterline. Draft marks provide a visual method for a rig operator or vessel to keep track of reserve buoyancy and to determine whether the vessel is sitting level in the water.

draw works (*n.*): The hoisting mechanism on a drilling rig. It is essentially a large winch that spools off or takes in the drilling line and raises or lowers the drill stem and bit.

drill (*v.*): To bore a hole in the earth, usually to find and remove subsurface formation fluids such as oil and gas.

driller (*n.*): The employee directly in charge of a drilling or work-over rig and crew. The driller's main duty is operation of the drilling and hoisting equipment, but this person is also responsible for downhole condition of the well, operation of downhole tools, and pipe measurements.

driller's BOP control panel (*n.*): A series of controls on the rig floor that the driller manipulates to open and close the blowout preventers.

driller's console (*n.*): A metal cabinet on the rig floor containing the controls that the driller manipulates to operate various components of the drilling rig.

driller's report (*n.*): A record kept on the rig for each tour to show the footage drilled, tests made on drilling fluid, bit record, and other noteworthy items.

drilling fluid (*n.*): Circulating fluid, one function of which is to lift cuttings out of the wellbore and to the surface. It also serves to cool the bit and to counteract downhole formation pressure. Although a mixture of barite, clay, water, and other chemical additives is the most common drilling fluid, wells can also be drilled by using air, gas, water, or oil-base mud as the drilling fluid. Also called circulating fluid and drilling mud. *See* mud.

drilling mud (*n.*): A specially compounded liquid circulated through the well-bore during rotary drilling and work-over operations. In addition to bringing cuttings to the surface, drilling mud cools and lubricates the bit and drill stem, protects against blowouts by holding back subsurface pressures, and deposits a mud cake on the wall of the borehole to prevent loss of fluids to the formation. Although it was originally a suspension of earth solids (especially clays) in water, the mud used in modern drilling operations is a more complex, three-phase mixture of liquids, reactive solids, and inert solids. The liquid phase might be fresh water, diesel oil, or crude oil and might contain one or more conditioners. *See* drilling fluid, mud.

drilling superintendent (*n.*): An employee, usually of a drilling contractor, who is in charge of all drilling operations that the contractor is engaged in. Also called a drilling foreman.

drill pipe (*n.*): Seamless steel or aluminum pipe made up in the drill stem between the top drive on the surface and the drill collars on the bottom. During drilling, it is usually rotated while drilling fluid is circulated through it. Drill pipe joints are available in three ranges of length: 18–22 ft. (5.49–6.71 m), 27–30 ft. (8.23–9.14 m), and 38–45 ft. (11.58–13.72 m). The most popular length is 27 to 30 ft.

dual-activity derrick (*n.*): A derrick so constructed that while one part of the derrick is being used for one drilling operation, another part can simultaneously be used for a different operation.

dual completion (*n.*): A single well that produces from two separate formations at the same time. Production from each zone is segregated by running two tubing strings with packers inside the single string of production casing, or by running one tubing string with a packer through one zone while the other is produced through the annulus. In a miniaturized dual completion, two separate 4.5 in. (11.4 cm) or smaller casing strings are run and cemented in the same wellbore.

duplex pump (*n.*): A reciprocating pump with two pistons or plungers used extensively as a mud pump on drilling rigs.

dynamic positioning (*n.*): A method by which a drilling rig or vessel floating offshore is maintained in position over an offshore well location without the use of mooring anchors. Generally several propulsion units, called thrusters, are on the hulls of the structure and are actuated by a sensing system. A computer to which the system feeds signals directs the thrusters to maintain the rig on location.

dynamic-positioning operator (*n.*): An employee on a drill ship or semisubmersible drilling rig whose primary duty is to monitor, operate, and maintain the equipment that maintains the rig on station while drilling.

E and P (*abbr.*): 1. Exploration and production. 2. Those activities that include subsurface studies, seismic and geophysical activities, locating underground hydrocarbon deposits, drilling for hydrocarbon deposits and bringing hydrocarbons to the surface, well completion, and field processing of hydrocarbons prior to entering the pipeline. 3. The upstream end of the petroleum industry.

elbow (*n.*): A fitting that allows two pipes to be joined at an angle of less than 180°, usually 90° or 45°.

electrician (*n.*): The rig crew member who maintains and repairs the electrical generation and distribution system on the rig.

elevator links (*n., pl.*): Cylindrical bars that support the elevators and attach them to the hook. Also called elevator bails.

elevators (*n., pl.*): On conventional rotary rigs and top-drive rigs, hinged steel devices with manual operating handles that crew members latch on to a tool joint (or a sub). Since the elevators are directly connected to the traveling block, or to the integrated traveling block in the top drive, when the driller raises or lowers the block or the top-drive unit, the drill pipe is also raised or lowered.

exchanger (*n.*): A piping arrangement that permits heat from one fluid to be transferred to another fluid as they travel countercurrently to one another.

expansion joint (*n.*): A device used to connect long lines of pipe to allow the pipe joints to expand or contract as temperature rises or falls.

exploitation well (*n.*): A well drilled to permit more-effective extraction of oil from a reservoir. Sometimes called a development well.

exploration (*n.*): The search for reservoirs of oil and gas, including aerial and geophysical surveys, geological studies, core testing, and drilling of wildcat wells

exploration agreement (*n.*): A contract between the mineral owner of a tract of land under which petroleum deposits lie or are suspected to lie, and a company or companies—such as an oil company or companies—that agree to develop the prospect. The company may carry out surface geological studies, seismic exploration, and eventually drill exploration wells.

exploration and production (*n.*): *See* E and P.

exploration geologists (*n., pl.*): Scientists who search for reservoirs of oil and gas and conduct geophysical surveys, geological studies, and core testing.

exploration well (*n.*): A well drilled either in search of an as-yet-undiscovered pool of oil or gas (a wildcat well) or to greatly extend the limits of a known pool. It involves a relatively high degree of risk. Exploratory wells may be classified as (1) wildcat, drilled in an unproven area, (2) field extension or step-out, drilled in an unproven area to extend the proven limits of a field, or (3) deep test, drilled within a field area but to unproven, deeper zones.

exploratory drilling (*n.*): Drilling involved with the initial discovery of hydrocarbons. Also called wildcatting.

exploratory well (*n.*): *See* exploration well.

fail-safe (*adj.*): Capable of compensating automatically and safely for a failure, such as of a mechanism or power source.

farm-out (*n.*): An agreement whereby the owner of a lease who does not wish to drill at the time agrees to assign the leasehold interest, or some part of it, to a third party who does wish to drill, conditional on the third party's drilling a well within the expiration date of the primary term of the lease.

The assignment may include the entire interest together with dry-hole money, or partial interest, or entire interest with or without an override. If an override is retained, the owner of the lease may retain an option to convert such overriding royalty retained to an agreed-upon working interest. A farm-out is distinguished from a joint-operating agreement by the fact that the partner farming out does not incur any of the drilling costs. The primary characteristic of a farm-out is the obligation of the third party to drill one or more wells on the farm-out acreage as a condition prerequisite to completion of the transfer of title to such third party.

farm out (*v.*): For a lessee, to agree to assign a leasehold interest to a third party, subject to stipulated conditions. *See* farm-out.

fast line (*n.*): The end of the drilling line that is affixed to the drum or reel of the draw works, so called because it travels with greater velocity than any other portion of the line. *Compare* deadline.

field (*n.*): A geographical area in which a number of oil or gas wells produce from a continuous reservoir. A field may refer to surface area only, or to underground productive formations as well. A single field may have several separate reservoirs at varying depths.

filter cake (*n.*): 1. Compacted solid or semisolid material remaining on a filter after pressure filtration of mud with a standard filter press. Thickness of the cake is reported in thirty-seconds of an inch or in millimeters. 2. The layer of concentrated solids from the drilling mud or cement slurry that forms on the walls of the borehole opposite permeable formations; also called wall cake or mud cake.

fingerboard (*n.*): A rack that supports the tops of the stands of pipe being stacked in the derrick or mast. It has several steel, fingerlike projections that form a series of slots into which the driller can place a stand of drill pipe after it is pulled out of the hole and removed from the drill string.

fish (*n.*): An object that is left in the wellbore during drilling or work-over operations and must be recovered before work can proceed. It can be anything from a piece of scrap metal to a part of the drill stem. (*v.*): 1. To recover from a well any equipment left there during drilling operations, such as a lost bit, or drill collar, or part of the drill string. 2. To remove from an older well certain pieces of equipment (such as packers, liners, or screen liner) to allow reconditioning of the well.

fixed-bore ram block (*n.*): In a ram blowout preventer, a steel block with elastomer surfaces that seal around drill pipe of a specific size. Besides forming a seal around drill pipe to confine well pressure in the annulus, they also support the load of the drill string—that is, the drill string can be hung off on them. *Compare* variable-bore ram block. *See* hang off, pipe ram.

flexible hose (*n.*): A type of tube or pipe that is bendable (flexible) so that repeated movements do not cause it to break; frequently used in subsea blowout preventer systems to conduct operating fluid from the accumulator on the surface to operating devices on the marine riser pipe.

flexible joint (flex joint) (*n.*): On floating offshore drilling rigs using a subsea blowout preventer, a device mounted between the annular preventer and the riser adapter on the lower marine riser package (LMRP). Flex joints bend laterally to prevent excessive bending forces from being exerted on the marine riser and the LMRP and BOP components (a bending moment is a force that lateral movement creates on an object). A flex joint typically allows 10° of offset from vertical. The riser adapter can be connected to the top of the flex joint's neck by a flange, a hub, or welding. *Compare* ball joint.

floating production and system offloader (FPSO) (*n.*): A floating offshore oil production vessel that has facilities for producing, treating, and storing oil from several producing wells and that puts (offloads) the treated oil into a tanker ship for transport to refineries on land. Some FPSOs are also capable of drilling, in which case they are termed floating production, drilling, and system offloaders (FPDSOs).

flowmeter (*n.*): An instrument that monitors, measures, or records the amount of fluid moving through a pipe or other container.

flow rate (q) (*n.*): 1. Time required for a given quantity of material to move a measured distance. 2. Weight or volume of material flowing per unit time. Also known as rate of flow.

flow-rate sensor (*n.*): A device mounted in the mud return flow line that detects the speed (flow rate) of the mudflow and sends a signal to an instrument on the rig floor and other rig locations to warn of a change in the return flow rate. An increase in flow may indicate that the well has kicked, while a decrease may indicate a loss of returns.

formation (*n.*): A bed or deposit composed throughout of substantially the same kind of rock, often a lithologic unit. Each formation is given a name, frequently as a result of the study of the formation outcrop at the surface and sometimes on the basis of fossils found in the formation.

formation boundary (*n.*): The horizontal limits of a formation.

formation breakdown (*n.*): The fracturing of a formation from excessive borehole pressure.

formation breakdown pressure (*n.*): The pressure at which a formation will fracture.

formation competency (*n.*): The ability of the formation to withstand applied pressure. Also called formation integrity.

formation competency test (*n.*): A test used to determine the amount of pressure required to cause a formation to fracture.

formation damage (*n.*): The reduction of permeability in a reservoir rock caused by the invasion of drilling fluid and treating fluids to the section adjacent to the wellbore. It is often called skin damage, in which a formation will crack from pressure in the wellbore.

formation fracturing (*n.*): A method of stimulating production by opening new flow channels in the rock surrounding a production well. Often called a frac job. Under extremely high hydraulic pressure, a fluid (such as distillate, diesel fuel, crude oil, dilute hydrochloric acid, water, or kerosene) is pumped downward through production tubing or drill pipe and forced out below a packer, or between two packers. The pressure causes cracks to open in the formation, and the fluid penetrates the formation through the cracks. Sand grains, aluminum pellets, walnut shells, or similar materials (propping agents) are carried in suspension by the fluid into the cracks. When the pressure is released at the surface, the fracturing fluid returns to the well. The cracks partially close on the pellets, leaving channels for oil to flow around them to the well.

formation gas (*n.*): Gas initially produced from an underground reservoir.

formation integrity (*n.*): *See* formation competency.

formation pressure (*n.*): The force exerted by fluids in a formation, recorded in the hole at the level of the formation with the well shut in. Also called reservoir pressure or shut-in bottom-hole pressure.

formation resistivity (*n.*): A measure of the electrical resistance of fluids in a formation.

formation sensitivity (*n.*): The tendency of certain producing formations to react adversely to invading filtrates.

formation strength (*n.*): The ability of a formation to resist fracture from pressures created by fluids in a borehole.

formation testing (*n.*): The gathering of pressure data and fluid samples from a formation to determine its production potential before choosing a completion method. Formation-testing tools include formation testers and drill stem test tools.

forward (*adv.*): In the direction of the bow on a ship or an offshore drilling rig.

FOSV (*abbr.*): Full-opening safety valve.

FPDSO (*abbr.*): Floating production, drilling, and system offloader.

FPSO (*abbr.*): Floating production and system offloader.

fracture (*n.*): A crack or crevice in a formation, either natural or induced.

fracture gradient (*n.*): The pressure gradient (psi/ft.) at which a formation accepts whole fluid from the wellbore. Also called frac gradient.

freeboard (*n.*): The vertical distance between the waterline and the freeboard deck on a ship, boat, or floating offshore drilling rig.

freeboard deck (*n.*): The uppermost continuous deck on a ship or floating rig that has a permanent means of closing all openings to the sea.

gallon (gal.) (*n.*): A unit of measure of liquid capacity that equals 3.785 liters and has a volume of 231 cu. in. (0.00379 m^3). A gallon of water weighs 8.34 lbs. (3.8 kg) at 60°F (16°C). The imperial gallon, formerly used in Great Britain, equals approximately 1.2 US gallons.

galvanized steel (*n.*): Steel that is coated with zinc to protect it from rusting.

gantry (*n.*): A platform made to carry a traveling crane and supported by towers or side frames running on parallel tracks.

gas (*n.*): A compressible fluid that completely fills any container in which it is confined. Technically, a gas will not condense when it is compressed and cooled, because a gas can exist only above the critical temperature for its particular composition. Below the critical temperature this form of matter is known as a vapor, because liquid can exist and condensation can occur. Sometimes the terms "gas" and "vapor" are used interchangeably. The latter should be used for those streams in which condensation can occur and that originate from, or are in equilibrium with, a liquid phase.

gas-cut mud (*n.*): A drilling mud that contains entrained formation gas, giving the mud a characteristically fluffy texture. When entrained gas is not released before the fluid returns to the well, the weight or density of the fluid column is reduced. Because a large amount of gas in mud lowers its density, gas-cut mud must be treated to reduce the chance of a blowout.

gas drilling (*n.*): A method of drilling that uses natural gas as the drilling fluid.

gas expansion (*n.*): When oil and gas are found in the same reservoir under pressure, the drilling of a well into the reservoir releases the pressure, causing the gas to expand. The expanding gas drives the oil toward and up the wellbore. The expansive energy of the gas can be harnessed whether the gas is in solution or forming a cap above the oil.

gas field (*n.*): A district or area from which natural gas is produced.

gas flow (*n.*): In well control, the streaming (the flow) of gas from the well when a formation containing gas is penetrated and is not contained by the mud.

gas injection (*n.*): The injection of gas into a reservoir to maintain formation pressure by gas drive and to reduce the rate of decline of the original reservoir drive. One type of gas injection uses gas that does not mix (i.e., that is not miscible) with the oil. Examples include natural gas, nitrogen, and flue gas.

gas injection well (*n.*): *See* gas input well.

gas input well (*n.*): A well into which gas is injected for the purpose of maintaining or supplementing pressure in an oil reservoir. More commonly called a gas injection well.

gas reservoir (*n.*): A geological formation containing a single gaseous phase. When this phase is produced, the surface equipment may or may not contain condensed liquid, depending on the temperature, pressure, and composition of the single reservoir phase.

gas seep (*n.*): The escape of natural gas to the surface from a formation that contains natural gas. *See* seep.

gas well (*n.*): A well that primarily produces gas.

gas zone (*n.*): An area in a reservoir that is occupied by natural gas.

gate valve (*n.*): An opening-and-closing device (a valve) that employs a slab of metal (a gate) with a hole in it that is moved up or down within the valve's body. When the hole in the gate is positioned opposite the opening in the valve, fluid flows through the valve. When the solid part of the gate is positioned opposite the valve's opening, flow stops. Gate valves are not used to regulate the flow of fluids through them—that is, they are either fully open or fully closed.

gauging (*n.*): Determining the liquid level of a tank so that its volume can be calculated. Usually done by lowering a weighted, graduated steel tape through the tank roof and noting the level at which the oil surface cuts the tape when the weight gently touches the tank bottom. It can also be done by measuring the distance between liquid height and a reference point and subtracting the distance from the gauge height to determine liquid height.

gel (*n.*): A semisolid, jellylike state assumed by some colloidal dispersions at rest. When agitated, the gel converts to a fluid state. Also a nickname for bentonite. (*v.*): To take the form of a gel; to set.

gel cement (*n.*): Cement or cement slurry that has been modified by the addition of bentonite.

geothermal gradient (*n.*): The increase in the temperature of the earth with increasing depth. It averages about 1°F per 60 ft. (1°C per 18.3 m) but may be considerably higher or lower.

gimbal (*n.*): A mechanical frame that permits an object mounted in it to remain in a stationary or near-stationary position regardless of movement of the frame. Gimbals are often used offshore to counteract undesirable wave motion.

gland (*n.*): A device used to form a seal around a reciprocating or rotating rod (as in a pump) to prevent fluid leakage. Specifically, the movable part of a stuffing box by which the packing is compressed.

gland packing (*n.*): Material placed around a gland to effect a seal around a reciprocating or rotating rod.

global positioning system (GPS) (*n.*): A global navigation satellite system (GNSS) that uses a constellation of satellites that transmit precise microwave signals that allow the GPS receivers to determine their current location, time, and velocity.

glycol (*n.*): A group of compounds used to dehydrate gaseous or liquid hydrocarbons, or to inhibit the formation of hydrates. Glycol is also used in engine radiators as an antifreeze. Commonly used glycols are ethylene glycol, diethylene glycol, and triethylene glycol.

Gondwanaland (*n.*): The southern part of the supercontinent Pangaea, comprising the future land masses of South America, Africa, Antarctica, Australia, and India.

gooseneck (*n.*): 1. The curved connection between the rotary hose and the swivel or top drive. 2. Any curved length of pipe that serves as a connection between one conduit to another. *See* swivel, top drive.

gpm (*abbr.*): 1. Gallons per minute when referring to rate of flow. 2. Gallons per 1,000 cu. ft. when referring to natural gas in terms of chromatograph analysis or theoretical gallons.

gravel (*n.*): Sand or glass beads of uniform size and roundness used in gravel packing.

gravel pack (*n.*): A mass of very fine gravel placed around a slotted liner in a well. See *gravel packing*.

gravel-pack *v*: to place a slotted or perforated liner in a well and surround it with gravel. *See* gravel packing.

gravel packing (*n.*): A method of well completion in which a slotted or perforated liner, often wire-wrapped, is placed in the well and surrounded by gravel. If open hole, the well is sometimes enlarged by underreaming at the point where the gravel is packed. The mass of gravel excludes sand from the wellbore but allows continued production.

gross registered tonnage (GRT) (*n.*): The total volume of a ship's interior measured in tons (units of 100 cu. ft. or 2.83 m^3).

gross tonnage (*n.*): The interior capacity of a ship or a mobile offshore drilling unit. The capacity is expressed in tons, although the actual measurement is in volume, 1 ton being equivalent to 100 cu. ft. of volume. This rule holds for measuring ship capacity for US maritime purposes. All principal maritime governments have their own rules describing how tonnage is to be measured.

guardrail (*n.*): A railing for guarding against danger or trespass. On a drilling or work-over rig, for example, guardrails are used on the rig floor to prevent persons from falling; guardrails are also installed on the mud pits and other high areas where there is any danger of falling.

guide shoe (*n.*): 1. A short, heavy, cylindrical section of steel filled with concrete and rounded at the bottom that is placed at the end of the casing string. It prevents the casing from snagging on irregularities in the borehole as it is lowered. A passage through the center of the shoe allows drilling fluid to pass up into the casing while it is being lowered, and allows cement to pass out during cementing operations. Also called casing shoe. 2. A device, similar to a casing shoe, placed at the end of other tubular goods.

gusher (*n.*): An oil well that has come in with such great pressure that the oil jets out of the well like a geyser. In reality, a gusher is a blowout and is extremely wasteful of reservoir fluids and drive energy. In the early days of the oil industry, gushers were common and, many times, were the only indication that a large reservoir of oil and gas had been struck. *See* blowout.

gyrocompass (*n.*): A compass card linked to a gyroscope in a gyroscopic multishot instrument. Unlike a magnetic compass, a gyrocompass is not affected by the earth's magnetic field.

gyroscope (*n.*): A wheel or disk mounted to spin rapidly about one axis but free to rotate about one or both of two axes perpendicular to each other. The inertia of the spinning wheel tends to keep its axis pointed in one direction regardless of how the other axes are rotated.

H_2S (*form.*): Hydrogen sulfide.

half-life (*n.*): The amount of time needed for half of a quantity of radioactive substance to decay or transmute into a nonradioactive substance. Half-lives range from fractions of seconds to millions of years.

handrail (*n.*): A railing or pipe along a passageway or stair that serves as a support or guard.

hanger (*n.*): A device placed or hung in the casing below the blowout preventer stack to form a pressure-tight seal. Pressure is then applied to the blowout preventer stack to test it for leaks. Also called a hanger plug.

hanger-packer (*n.*): A device that fits around the top of a string of casing or a string of tubing and supports part of the weight of the casing or tubing while providing a pressure-tight seal.

hanger plug (*n.*): A device placed or hung in the casing below the blowout preventer stack to form a pressure-tight seal. Pressure is then applied to the blowout preventer stack to test it for leaks.

hang off (*v.*): To close a ram blowout preventer around the drill pipe when the annular preventer has previously been closed to offset the effect of heave on floating offshore rigs during well control procedures.

hatch (*n.*): 1. An opening in the roof of a tank through which a gauging line may be lowered to measure its contents. 2. The opening from the deck into the cargo space of ships.

heading (*n.*): 1. Intermittent flow of fluid from a well. 2. Direction in which a vessel is pointed.

heating coils (*n., pl.*): (marine) A system of piping in tank bottoms in which steam is carried as required to heat high-pour-point liquid cargoes to pumpable viscosity level.

heave compensator (*n.*): A device that moves with the heave of a floating offshore drilling rig to prevent the bit from being lifted off the bottom of the hole and then dropped back down (i.e., to maintain constant weight on the bit).

heaving (*n.*): The partial or complete collapse of the walls of a hole resulting from internal pressures due primarily to swelling from hydration or the formation of gas pressures. *See* caving.

heavyweight drill pipe (*n.*): Drill pipe that is like conventional drill pipe, except it is manufactured with a thicker wall, increasing its weight and tensile strength. For example, 5 in., IEU-grade, E75 standard-weight drill pipe weighs 19.5 lbs. per ft. (29 kg/m), while 5 in., IEU-grade, E75 heavyweight drill pipe weighs 25.60 lbs. per ft. (38.4 kg/m). Heavyweight drill pipe is placed in the top of the string to support a very long string of pipe. Heavyweight drill pipe may be used in the drill stem when high-tensile-strength drill pipe is required but high-grade steel cannot be used because of the high-grade steel's susceptibility to hydrogen embrittlement, which is a form of corrosion caused by hydrogen sulfide.

helideck (*n.*): The designated place for helicopter landings and departures. Sometimes called a heliport.

high-pressure wellhead (*n.*): In drilling from floating drilling rigs, a subsea device that is installed on top of the casing to provide a foundation for the blowout preventer stack to support and house subsequent casing strings, and to provide a seal and a locking arrangement between the surface casing and blowout preventer stack.

hinged master bushing (*n.*): A two-piece master bushing that has a jointed, swinging device (a hinge) on each half, into which large pins fit to hold the bushing together. A two-piece-insert bowl to hold the slips fits inside this type of master bushing.

hoist (*n.*): 1. An arrangement of pulleys and wire rope or chain used for lifting heavy objects; a winch or similar device. 2. The draw works. (*v.*): To raise or lift.

hoisting (*n.*): The process of lifting.

hoisting cable (*n.*): The cable that supports drill pipe, swivel, hook, and traveling block on a rotary drilling rig.

hoisting components (*n., pl.*): Draw works, drilling line, and traveling and crown blocks. Auxiliary hoisting components include catheads, cat shaft, and air hoist.

hole (*n.*): 1. In drilling operations, the wellbore or borehole. *See* borehole, wellbore.

hole opener (*n.*): A device used to enlarge the size of an existing borehole. It has teeth arranged on its outside circumference to cut the formation as it rotates.

holiday (*n.*): A gap or void in coating on a pipeline, or in paint on a metal surface.

hook (*n.*): A large, hook-shaped device from which the swivel is suspended. It is designed to carry maximum loads ranging from 100 to 650 tons (90 to 590 tonnes) and turns on bearings in its supporting housing. A strong spring within the assembly cushions the weight of a stand (90 ft., about 27 m) of drill pipe, thus permitting the pipe to be made up and broken out with less damage to the tool joint threads. Smaller hooks without the spring are used for handling tubing and sucker rods.

hook load (*n.*): The weight of the drill stem that is suspended from the hook.

horizontal drilling (*n.*): Deviation of the borehole at least 80° from vertical so that the borehole penetrates a productive formation in a manner parallel to the formation. A single horizontal hole can effectively drain a reservoir and eliminate the need for several vertical boreholes.

horizontal well (*n.*): A directional well that is drilled about 90° from vertical so that the wellbore runs parallel to the surface. Horizontal wells have proven to be an effective way to exploit some reservoirs—for example, in limestone reservoirs with vertical fractures that contain hydrocarbons, a horizontal wellbore can penetrate several of the fractures to efficiently produce them.

human-machine interface (HMI) (*n.*): The apparatus that allows a human operator to work and interact with nonhuman equipment.

hydrate (*n.*): A hydrocarbon and water compound that is formed under reduced temperature and pressure in places where gas that contains water vapor occurs. For example, hydrates can form in gathering, compression, and transmission facilities for gas. They can also form in deepwater drilling, where low temperatures occur on or near the seafloor and where gas containing water vapor may be encountered by the borehole. Hydrates can accumulate in troublesome amounts and impede fluid flow. They resemble snow or ice and decompose at atmospheric pressure. (*v.*): To enlarge by taking water on or in.

hydraulic actuator (*n.*): A cylinder or fluid motor that converts hydraulic power into useful mechanical work; mechanical motion produced may be linear, rotary, or oscillatory.

hydraulic fracturing (*n.*): An operation in which a specially blended liquid is pumped down a well and into a formation under pressure high enough to cause the formation to crack open, forming passages through which oil can flow into the wellbore. Sand grains, aluminum pellets, glass beads, or similar materials are carried in suspension into the fractures. When the pressure is released at the surface, the fractures partially close on the proppants, leaving channels for oil to flow through to the well.

hydraulic torque wrench (*n.*): A hydraulically powered device that can break out or make up tool joints and ensure accurate torque. It is fitted with a repeater gauge so that the driller can monitor tool joints as they go downhole, doubly ensuring that all have the correct torque.

hydrocarbons (*n., pl.*): Organic compounds of hydrogen and carbon whose densities, boiling points, and freezing points increase as their molecular weights increase. Although composed of only two elements, hydrocarbons exist in a variety of compounds because of the strong affinity of the carbon atom for other atoms and for itself. The smallest molecules of hydrocarbons are gaseous; the largest are solids. Petroleum is a mixture of many different hydrocarbons.

hydrogen sulfide (*n.*): A flammable, colorless, gaseous compound of hydrogen and sulfur (H_2S), which in small amounts has the odor of rotten eggs. Sometimes found in petroleum, it causes the foul smell of petroleum fractions. In dangerous concentrations it is extremely corrosive and poisonous, causing damage to skin, eyes, breathing passages, and lungs and attacking and paralyzing the nervous system, particularly the part controlling the lungs and heart. In large amounts the sense of smell is lost. Also called hepatic gas or sulfureted hydrogen.

hydrophone (*n.*): A device trailed in an array behind a boat in offshore seismic exploration that is used to detect sound reflections, convert them to electric current, and send them through a cable to recording equipment on the boat.

hydrostatic head (*n.*): The pressure exerted by a body of water at rest. The hydrostatic head of fresh water is 0.433 psi per foot of height. Other liquids may be determined by comparing their gravities with the gravity of water.

hydrostatic pressure (HP) (*n.*): The force exerted by a body of fluid at rest. It increases directly with the density and the depth of the fluid and is expressed in many different units, including lbs. / sq. in. or kilopascals (kPa). The hydrostatic pressure of fresh water is 0.433 lbs. / sq. in. per ft. (9.792 kPa/m) of depth. In drilling, the term refers to the pressure exerted by the drilling fluid in the wellbore. In a water drive field, the term refers to the pressure that may furnish the primary energy for production. *See* hydrostatic head.

hydrostatic testing (*n.*): The most common final quality control check of the structural soundness of a pipeline. The line is filled with water and then pressured to a designated point. This pressure is maintained for a specified period of time, and any ruptures or leaks revealed by the test are repaired. The test is repeated until no problems are noted.

hypothermia (*n.*): Reduced body temperature caused by overexposure to chilling temperatures.

IADC (*abbr.*): International Association of Drilling Contractors.

IBOP (*abbr.*): Inside blowout preventer.

impermeable (*adj.*): Preventing the passage of fluid. A formation may be porous yet impermeable if there is an absence of connecting passages between the voids within it. *See* permeability.

inclining experiment (*n.*): A process to determine the position of an offshore rig or a vessel's center of gravity. A weight having a known value is placed on board the vessel and is moved a measured distance perpendicular to the vessel's centerline plane. Then the resulting angle of list is measured to determine the vessel's center of gravity.

initial stability (*n.*): In a ship or offshore drilling rig, the condition in which the vessel maintains equilibrium (is stable) when its metacentric height is above its center of gravity.

insert (*n.*): 1. A cylindrical object, rounded, blunt, or chisel-shaped on one end and usually made of tungsten carbide that is inserted in the cones of a bit, the cutters of a reamer, or the blades of a stabilizer to form the cutting element of the bit or the reamer, or the wear surface of the stabilizer. Also called a compact. 2. A removable part molded to be set into the opening of the master bushing so that various sizes of slips may be accommodated. Also called a bowl. 3. A removable, hard-steel, serrated piece that fits into the jaws of the tongs and firmly grips the body of the drill pipe or drill collars while the tongs are making up or breaking out the pipe. Also called die.

inside blowout preventer (IBOP) (*n.*): Any one of several types of valve installed in the drill stem or in a top drive to prevent high-pressure fluids from flowing up the drill stem and into the atmosphere. Also called an internal blowout preventer.

inside BOP (*abbr.*): Inside blowout preventer.

intact stability (*n.*): For offshore rigs and ships, the condition in which a vessel is in a stable state—that is, the vessel is maintaining equilibrium.

intermediate casing string (*n.*): The string of casing set in a well after the surface casing but before production casing is set, to keep the hole from caving and to seal off troublesome formations. In deep wells, one or more intermediate strings may be required. Sometimes called protection casing.

International Association of Classification Societies (IACS) (*n.*): An organization formed in 1968 and headquartered in London, England, it is dedicated to safe ships and clean seas. IACS provides technical support, compliance verification, and research and development in the interests of maritime safety and regulation. IACS 's classification design, construction, and through-life compliance rules and standards cover more than 90% of the world's cargo-carrying tonnage.

International Association of Drilling Contractors (IADC) (*n.*): An organization of drilling contractors that sponsors or conducts research on education, accident prevention, drilling technology, and other matters of interest to drilling contractors and their employees.

International Maritime Organization (IMO) (*n.*): An agency of the United Nations that provides an international forum for nations to discuss international cooperative efforts to improve marine safety and to protect the ocean environment. International treaties on safety (SOL AS) and environmental protection (MARPOL) have been negotiated at IMO. The governing body of the IMO is the assembly, which meets once every two years and is open to all 136 member states (countries).

Iron Roughneck™ (*n.*): A manufacturer's name for a floor-mounted combination of a spinning wrench and a torque wrench. The Iron Roughneck moves into position hydraulically and eliminates the manual handling involved with suspended individual tools.

jacket water (*n.*): Water that fills, or is circulated through, a housing that partially or wholly surrounds a vessel or machine to remove, add, or distribute heat, thereby controlling the temperature within the vessel or machine.

jack-up drilling rig (*n.*): A mobile, bottom-supported offshore drilling structure with columnar or open-truss legs that support the deck and hull. When positioned over the drilling site, the bottoms of the legs penetrate the seafloor. A jack-up rig is towed or propelled to a location with its legs up. Once the legs are firmly positioned on the bottom, the deck and hull height are adjusted and leveled. Also called self-elevating drilling unit.

jet (*n.*): 1. A hydraulic device operated by a centrifugal pump used to clean the mud pits, or tanks, and to mix mud components. 2. In a perforating gun using shaped charges, a highly penetrating, fast-moving stream of exploded particles that forms a hole in the casing, cement, and formation.

jet bit (*n.*): A drilling bit having replaceable nozzles through which the drilling fluid is directed in a high-velocity stream to the bottom of the hole to improve the efficiency of the bit. *See* bit.

joint (*n.*): In drilling, a single length (from 16 to 45 ft., or 5 to 14.5 m, depending on its range length) of drill pipe, drill collar, casing, or tubing that has threaded connections at both ends. Several joints screwed together constitute a stand of pipe.

junk (*n.*): Metal debris lost in a hole. Junk may be a lost bit, pieces of a bit, milled pieces of pipe, wrenches, or any relatively small object that impedes drilling or completion and must be fished out of the hole. (*v.*):To abandon (as a nonproductive well).

junk basket (*n.*): A device made up on the bottom of the drill stem or on wire line to catch pieces of junk from the bottom of the hole. Circulating the mud or reeling in the wire line forces the junk into a barrel in the tool, where it is caught and held. When the basket is brought back to the surface, the junk is removed. Also called a junk sub or junk catcher.

keel (K) (*n.*): A centerline strength tube running fore and aft along the bottom of a ship or a floating offshore drilling rig and forming the backbone of the structure.

keeper (*n.*): 1. An exploration well intended for completion. 2. A device or latch that locks something in place. 3. A soft iron bar placed between the poles of a permanent magnet when it is not being used. A keeper protects the magnet's poles from being demagnetized if the magnet is dropped or struck with a hard object.

kelly (*n.*): On drilling rigs that do not use a top drive to rotate the bit, a heavy steel tubular device, four or six sided, suspended from the swivel through the rotary table and connected to the top joint of drill pipe to turn the drill stem as the rotary table turns. It has a bored passageway that permits fluid to be circulated into the drill stem and up the annulus, or vice versa. Kellys manufactured to API specifications are available in four- or six-sided versions, are either 40 or 54 ft. (12 or 16 m) long, and have diameters as small as 2.5 in. (6 cm) and as large as 6 in. (15 cm).

kelly bushing (KB) (*n.*): A special device placed around the kelly that mates with the kelly flats and fits into the master bushing of the rotary table. The kelly bushing is designed so that the kelly is free to move up or down through it. The bottom of the bushing may be shaped to fit the opening in the master bushing, or it may have pins that fit into the master bushing. In either case, when the kelly bushing is inserted into the master bushing and the master bushing is turned, the kelly bushing also turns. Since the kelly bushing fits on to the kelly, the kelly turns, and since the kelly is made up to the drill stem, the drill stem turns. Also called the drive bushing.

key seat (*n.*): 1. An under-gauge channel or groove cut in the side of the borehole and parallel to the axis of the hole. A key seat results from the rotation of pipe on a sharp bend in the hole. 2. A groove cut parallel to the axis in a shaft or a pulley bore.

kick (*n.*): An entry of water, gas, oil, or other formation fluid into the wellbore during drilling, work-over, or other operations. It occurs because the pressure exerted by the column of drilling or other fluid in the wellbore is not great enough to overcome the pressure exerted by the fluids in a formation exposed to the wellbore. If prompt action is not taken to control the kick, or kill the well, a blowout may occur. *See* blowout.

kicked off (*v.*): When a well is deviated in another direction. *See* deviating, deviation, kickoff point.

kick off (*v.*): 1. To bring a well into production; used most often when gas is injected into a gas lift well to start production. 2. In work-over operations, to swab a well to restore it to production. 3. To deviate a wellbore from the vertical, as in directional drilling.

kickoff point (KOP) (*n.*): The depth in a vertical hole at which a deviated or slant hole is started; used in directional drilling.

kill (*v.*): 1. In drilling, to control a kick by taking suitable preventive measures (e.g., to shut in the well with the blowout preventers, circulate the kick out, and increase the weight of the drilling mud). 2. In production, to stop a well from producing oil and gas so that reconditioning of the well can proceed. Production is stopped by circulating a kill fluid into the hole.

kill fluid (*n.*): Mud or other fluid in a wellbore whose weight, or density, creates pressure great enough to equal or exceed the pressure exerted by formation fluids.

kill line (*n.*): A pipe, or conduit, usually attached to openings in the blowout preventer stack below the ram blowout preventers, that allows drilling mud or other fluid to be pumped into the well to control a well when it is not possible to pump down the drill string.

kill mud (*n.*): *See* kill fluid.

kill sheet (*n.*): A printed form that contains blank spaces for recording information about killing a well. It is provided to remind personnel of the necessary steps to take to kill a well.

kink (*n.*): A loop in a wire rope that, having been pulled tight, causes permanent distortion.

knot (kn., kt.) (*n.*): A unit of speed equal to 1 nautical mile (1.852 km or about 1⅐ statute miles) per hour.

land a wellhead (*v.*): To attach casing heads and other wellhead equipment not already in place at the time of well completion.

land casing (*v.*): To install casing so that it is supported in the casing head by slips.

landing string (*n.*): Offshore, to land the wellhead on casing beneath the water on the seafloor, it is run in a landing string (drill pipe or tubing) to the seafloor or inside casing.

leak-off point (*n.*): In a leak-off test, the pressure at which drilling mud begins to leak off, or enter, the formation from the borehole.

lease (*n.*): 1. A legal document executed between a landowner, as lessor, and a company or individual, as lessee, that grants the right to exploit the premises for minerals or other products; the instrument that creates a leasehold or working interest in minerals. 2. The area where production wells, stock tanks, separators, LACT units, and other production equipment are located.

lee (*n.*): The side of a ship, vessel, or floating offshore rig away from the source of the wind. For example, if the wind is blowing from the north, the lee is on the south side of the ship, vessel, or rig. *Compare* windward.

leeward (*adj.*): The side of a rig that is away from the effects of wind and waves. Also called downwind.

lifeboat (*n.*): A small boat hoisted on davits or carried on one of the upper decks of a vessel that can be quickly lowered into the water in case of an emergency.

light displacement (*n.*): On mobile offshore drilling rigs, the weight of the rig with all permanently attached equipment but without fuel, supplies, crew, ballast, drill pipe, and so forth.

limber hole (*n.*): A hole cut in a structural member of a ship or offshore drilling rig, usually in a tank, to allow water to pass through freely.

liner (*n.*): 1. A string of pipe used to case an open hole below the existing casing. A liner extends from the setting depth up into another string of casing, usually overlapping about 100 ft. (30.5 m) above the lower end of the intermediate or the oil string. Liners are nearly always suspended from the upper string by a hanger device. 2. A relatively short length of pipe with holes or slots that is placed opposite a producing formation. Usually such liners are wrapped with specially shaped wire designed to prevent the entry of loose sand into the well as it is produced. They are also often used with a gravel pack. 3. In jet perforation guns, a conically shaped metallic piece that is part of a shaped charge. It increases the efficiency of the charge by increasing the penetrating ability of the jet. *See* jet. 4. A replaceable tube that fits inside the cylinder of an engine or a pump.

liquefied natural gas (LNG) (*n.*): A liquid composed chiefly of natural gas (i.e., mostly methane). Natural gas is liquefied to make it easy to transport if a pipeline is not feasible (such as across a body of water). Not as easily liquefied as LPG, LNG must be put under low temperature and high pressure or under extremely low (cryogenic) temperature and close to atmospheric pressure to liquefy.

list (*n.*): The position of a ship or offshore drilling rig that heels to one side because of a shift in cargo, machinery, or supplies.

LMRP (*abbr.*): *See* lower marine-riser package.

LMRP connector (*n.*): Used to remotely connect the BOP assembly to the wellhead, and the LMRP to the BOP.

load line (*n.*): A line, painted or cut on the outside of a floating rig or ship's hull, that marks the maximum waterline when the rig or ship is loaded with the greatest amount of cargo that it can safely carry.

log a well (*v.*): To run any of the various logs used to ascertain downhole information about a well.

logbook (*n.*): A book used by station engineers, dispatchers, and gaugers for keeping notes on current operating data.

logging while drilling (LWD) (*n.*): Logging measurements obtained by measurement while drilling techniques as the well is being drilled.

lower marine-riser package (LMRP) (*n.*): 1. The equipment that attaches the bottom part of the marine riser to the subsea blowout preventer (BOP) stack. It includes the BOP connector, a flexible joint to compensate for side-to-side movement, and the marine riser connector. 2. A hydraulically actuated device that is used to remotely connect the blowout preventer stack to the wellhead and the lower marine riser package to the blowout preventer stack.

magnetic survey (*n.*): An exploration method in which an instrument that measures the intensity of the natural magnetic forces existing in the earth's subsurface is passed over the surface or through the water. The instrumentation detects deviations in magnetic forces, and such deviations may indicate the existence of underground formations that favor the entrapment of hydrocarbons.

main deck (*n.*): The principal continuous deck of a ship or offshore drilling rig, running from fore to aft, from which the freeboard is determined.

make a trip (*v.*): To hoist the drill stem out of the wellbore to perform one of a number of operations, such as changing bits or taking a core, and then returning the drill stem to the wellbore.

make hole (*v.*): To deepen the hole made by the bit (i.e., to drill ahead; to run casing or pipe).

make up (*v.*): 1. To assemble and join parts to form a complete unit (e.g., to make up a string of drill pipe). 2. To screw together two threaded pieces. *Compare* break out. 3. To mix or prepare (e.g., to make up a tank of mud). 4. To compensate for (e.g., to make up for lost time).

managed pressure drilling (MPD) (*n.*): A type of drilling that precisely controls the annular pressure that occurs when drilling. In MPD, the amount of annular pressure that is maintained on the wellbore is directly related to the formations through which the borehole passes.

materials coordinator (*n.*): A person who oversees the acquisition of supplies, equipment, and other resources for a company or organization. This person's duties may also include managing and controlling warehouse operations, ordering materials, and managing inventory. Also called materials man, warehouse supervisor.

mean draft (*n.*): A measurement of draft resulting from the average of several draft readings taken between the bow and stern of a floating vessel. If the vessel has a straight keel, mean draft occurs at the midpoint of the waterline length.

measured depth (MD) (*n.*): The total length of the wellbore, measured in feet along its actual course through the earth. Measured depth can differ from true vertical depth, especially in directionally drilled wellbores.

measurement while drilling (MWD) (*n.*): 1. Directional and other surveying during routine drilling operations to determine the angle and direction by which the wellbore deviates from the vertical. 2. Any system of measuring downhole conditions during routine drilling operations.

metric ton (*n.*): A measurement equal to 1,000 kg or 2,204.6 avoirdupois. In some oil-producing countries, production is reported in metric tons. One metric ton is equivalent to about 7.4 barrels (bbls.) (42 US gal. = 1 bbl.) of crude oil with a specific gravity of 0.0184, or 36° API. In the SI system it is called a tonne.

mill (*n.*): A downhole tool with rough, sharp, extremely hard cutting surfaces for removing metal, packers, cement, sand, or scale by grinding or cutting. Mills are run on drill pipe or tubing to grind up debris in the hole, remove stuck portions of drill stem or sections of casing for sidetracking, and ream out tight spots in the casing. They are also called junk mills, reaming mills, and so forth, depending on their use. (*v.*): To use a mill to cut or grind metal objects that must be removed from a well.

mixing mud (*n.*): Preparation of drilling fluids from a mixture of water and other fluids and one or more of the various dry mud-making materials such as clay and chemicals.

mixing tank (*n.*): Any tank or vessel used to mix components of a substance (as in the mixing of additives with drilling mud).

mobile offshore drilling unit (MODU) (*n.*): A drilling rig that drills offshore exploration and development wells. It floats on the surface of the water when being moved from one well site to another, but it may or may not float once drilling begins. Two basic types of mobile offshore drilling units are used to drill most offshore wildcat wells: bottom-supported offshore drilling rigs and floating drilling rigs. Bottom-supported units include jack-ups, and floating rigs include semisubmersibles and drillships.

monkeyboard (*n.*): The working platform where pipe or tubing (when stored vertically) is accessible and can be as high as 90 ft. (27 m) or higher in the derrick or mast. The monkeyboard provides a small platform to raise the derrick hand to the proper height for handling the top of the pipe if needed.

moon pool (*n.*): A walled hole or well in the hull of a drillship, ship-shaped barge, or semisubmersible drilling rig (usually in the center) through which the drilling assembly and other assemblies pass while a well is being drilled, completed, or abandoned.

motion compensator (*n.*): Any device (such as a bumper sub or heave compensator) that serves to maintain constant weight on the bit in spite of vertical motion of a floating offshore drilling rig.

mousehole (*n.*): An opening in the rig floor, usually lined with pipe, into which a length of drill pipe is placed temporarily for later connection to the drill string.

mud (*n.*): The liquid circulated through the wellbore during rotary drilling and work-over operations. In addition to its function of bringing cuttings to the surface, drilling mud cools and lubricates the bit and the drill stem, protects against blowouts by holding back subsurface pressures, and deposits a mud cake on the wall of the borehole to prevent loss of fluids to the formation. Although it originally was a suspension of earth solids (especially clays) in water, the mud used in modern drilling operations is a more complex, three-phase mixture of liquids, reactive solids, and inert solids. The liquid phase may be fresh water, diesel oil, or crude oil and may contain one or more conditioners.

mud booster line (*n.*): In drilling from floating offshore drilling rigs that use a subsea blowout preventer stack, a line (pipe) sometimes provided on riser joints to increase the return velocity of the mud in the riser. It helps move heavy cuttings up the riser when normal pump circulation is inadequate. The mud booster line is connected to a pump on the rig that enables the driller to circulate drilling fluid from the bottom of the riser.

mud cake (*n.*): The sheath of mud solids that forms on the wall of the hole when liquid from mud filters into the formation. Also called filter cake or wall cake.

mud engineer (*n.*): An employee of a drilling-fluid supply company whose duty it is to test and maintain the drilling-mud properties that are specified by the operator.

mud-gas separator (*n.*): A device that removes gas from the mud coming out of a well when a kick is being circulated out, or when the well is being drilled underbalanced and gas must be removed from the liquid part of the drilling fluid.

mud pump (*n.*): A large, high-pressure reciprocating pump used to circulate the mud on a drilling rig. A typical mud pump is a single- or double-acting, two- or three-cylinder piston pump whose pistons travel in replaceable liners and are driven by a crankshaft actuated by a motor.

mud tank (*n.*): One of a series of open tanks, usually made of steel plate, through which the drilling mud is cycled to remove sand and fine sediments. Additives are mixed with the mud in the tanks, and the fluid is temporarily stored there before being pumped back into the well. Modern rotary drilling rigs are generally provided with three or more tanks fitted with built-in piping, valves, and mud agitators. Also called mud pits.

multipay zone (*n.*): Two or more hydrocarbon-producing formations that are penetrated by a single wellbore.

multiplex electronic control (MUX) system (*n.*): An arrangement of equipment employed on floating rigs drilling in water depths greater than 5,000 ft. (1,500 m) to overcome delays in the transmission of signals to close and open the subsea blowout preventers. Surface electronics transmit electronic command signals through cable reels to subsea multiplex electronics packages that decode and deliver the commands to solenoid valves.

natural gas (*n.*): A highly compressible, highly expansible mixture of hydrocarbons with a low specific gravity occurring naturally in a gaseous form. Besides containing hydrocarbon gases, natural gas may contain appreciable quantities of nitrogen, helium, carbon dioxide, hydrogen sulfide, and water vapor. Although gaseous at normal temperatures and pressures, the gases making up the mixture that is natural gas are variable in form and may be found either as gases or as liquids under suitable conditions of temperature and pressure.

natural-gas liquids (NGL) (*n., pl.*): Those hydrocarbons liquefied at the surface in field facilities or in gas-processing plants. Natural-gas liquids include propane, butane, and natural gasoline.

nautical mile (*n.*): A unit of length used in sea and air navigation that is based on the length of 1 minute of arc of a great circle; unit equal to 1,852 m (about 6,076 ft.).

net tonnage (*n.*): The gross tonnage of a ship or a mobile offshore drilling rig less all spaces that are not or cannot be used for carrying cargo, expressed in tons equal to 100 cu. ft.

nominal strength (*n.*): Wire rope strength that the manufacturer calculates using a standard procedure established by the wire rope industry. Also called catalog strength.

nominal volume (*n.*): The quantity assigned to a tank or vessel for the purpose of identification only; the exact volume may be somewhat different from the nominal volume.

nondestructive testing (NDT) (*n.*): Testing designed to evaluate the quality both of production and field welds without altering their basic properties or affecting their future usefulness. The most-common nondestructive testing is radiographic, or x-ray, testing.

normal formation pressure (*n.*): Formation fluid pressure equivalent to about 0.465 lbs. per sq. in. per ft. (10.5 kPa/m) of depth from the surface. If the formation pressure is 4,650 lbs. / sq. in. (32,062 kPa) at 10,000 ft. (3,048 m), it is considered normal developed by a column of fluid, since the depth of the column increases when the column contains a fluid of normal density. This gradient varies from area to area, but along the Gulf Coast of the United States it is considered to be 0.465–0.468 psi/ft. (10.53–10.59 kPa/m), which is the pressure developed by the salt water that naturally occurs in the formations of that area.

nozzle (*n.*): 1. A passageway through jet bits that causes the drilling fluid to be ejected from the bit at high velocity. The jets of mud clear the bottom of the hole. Nozzles come in different sizes that can be interchanged on the bit to adjust the velocity with which the mud exits the bit. 2. The part of the fuel system of an engine that has small holes in it to permit fuel to enter the cylinder. Properly known as a fuel injection nozzle but also called a spray valve. The needle valve is directly above the nozzle.

odorant (*n.*): A chemical, usually a mercaptan, that is added to natural gas so that the presence of the gas can be detected by the smell.

off-center wear (*n.*): A type of bit wear in which the cutters on the cones wear in an uneven pattern because of the whirling action of the bit as it drills. Whirling is the motion a bit makes when it does not rotate around the center; instead it drills with a spiral motion.

offshore drilling (*n.*): Drilling for oil or gas in an ocean, gulf, or sea, usually on the Outer Continental Shelf. A drilling unit for offshore operations may be a mobile floating vessel with a ship or barge hull, a semisubmersible or submersible base, a self-propelled or towed structure with jacking legs (jack-up drilling rig), or a permanent structure used as a production platform when drilling is completed. Generally wildcat wells are drilled from mobile floating vessels or from jack-ups, while development wells are drilled from platforms or jack-ups.

offshore installation manager (OIM) (*n.*): A qualified and certified person with marine and drilling knowledge who is in charge of all operations on a MO DU.

offshore rig (*n.*): Any of various types of drilling structures designed for use in drilling wells in oceans, seas, bays, gulfs, and so forth. Offshore rigs include platforms, jack-up drilling rigs, semisubmersible drilling rigs, submersible drilling rigs, and drill ships.

oil-and-gas separator (*n.*): An item of production equipment used to separate liquid components of the well fluid from gaseous elements. Separators are either vertical or horizontal and either cylindrical or spherical. Separation is accomplished principally by gravity, with the heavier liquids falling to the bottom and the gas rising to the top. A float valve or other liquid-level control regulates the level of oil in the bottom of the separator.

oil-base mud (*n.*): A drilling or work-over fluid in which oil is the continuous phase and that contains from less than 2% and up to 5% water. This water is spread out, or dispersed, in the oil as small droplets.

oil sand (*n.*): 1. A sandstone that yields oil. 2. (by extension) Any reservoir that yields oil, whether or not it is sandstone.

oil scout (*n.*): A person who gathers data on new oil and gas wells and other industry developments.

oil seep (*n.*): A surface location where oil appears, the oil having permeated its subsurface boundaries and accumulated in small pools or rivulets. Also called an oil spring.

operational draft (*n.*): On a semisubmersible drilling rig, the draft that yields the best motion characteristics when the rig is in drilling mode with a maximum deck load. *See* draft.

outer barrel (*n.*): In the telescopic joint of a marine riser assembly, a pipe that attaches to the top joint of the marine riser assembly and to which are attached the riser tensioner lines. The inner barrel of the telescopic joint moves up and down within the outer barrel as the floating offshore rig heaves.

Outer Continental Shelf (OCS) (*n.*): The land seaward from areas subject to state mineral ownership to a distance of roughly 200 nautical miles. Boundaries of the OCS are set by law.

packer (*n.*): A piece of downhole equipment that consists of a sealing device, a holding or setting device, and an inside passage for fluids. It is used to block the flow of fluids through the annular space between pipe and the wall of the wellbore, by sealing off the space between them. In production it is usually made up in the tubing string some distance above the producing zone. A packing element expands to prevent fluid flow except through the packer and tubing. Packers are classified according to configuration, use, and method of setting, and whether or not they are retrievable (that is, whether they can be removed when necessary, or whether they must be milled or drilled out and thus destroyed).

pad eye (*n.*): A steel plate with an opening (the eye) that is usually welded to a heavy object and to which a sling or other lifting line is attached to facilitate the object being moved, lifted, or lowered.

payload (*n.*): The load carried by a ship or floating offshore rig exclusive of that necessary for its operation.

pay sand (*n.*): The producing formation, often one that is not even sandstone. Also called pay, pay zone, and producing zone.

perforate (*v.*): To pierce the casing wall and cement of a wellbore to provide holes through which formation fluids may enter, or to provide holes in the casing so that materials may be introduced into the annulus between the casing and the wall of the borehole. Perforating is accomplished by lowering into the well a perforating gun, or perforator, which fires electrically detonated bullets or shaped charges.

pipe dolly (*n.*): Any device equipped with rollers and used to move drill pipe or collars. It is usually placed under one end of the pipe while the pipe is being lifted from the other end by an air hoist line.

pipe handler (*n.*): In a top drive, the power and spinning wrenches built into the unit that spins, makes up, breaks out, and backs up the pipe.

pipe-racker machines (*n., pl.*): Mechanized devices used to rack pipe.

pit gain (*n.*): An increase in the average level of mud maintained in each of the mud pits, or tanks. If no mud or other substances have been added to the mud circulating in the well, then a pit gain is an indication that formation fluids have entered the well and that a kick has occurred.

pit level (*n.*): Height of drilling mud in the mud tanks, or pits.

pit-level indicator (*n.*): One of a series of devices that continuously monitor the level of the drilling mud in the mud tanks. The indicator usually consists of float devices in the mud tanks that sense the mud level and transmit data to a recording and alarm device (a pit volume recorder) mounted near the driller's position on the rig floor. If the mud level drops too low or rises too high, the alarm sounds to warn the driller of lost circulation or to prevent a blowout.

Pit Volume Totalizer™ (PVT) (*n.*): Trade name for a type of pit level indicator. *See* pit-level indicator.

Plimsoll mark (*n.*): A mark placed on the side of a floating offshore drilling rig or ship to denote the maximum depth to which it may be loaded or ballasted. The line is set in accordance with local and international rules for safety of life at sea (SOL AS).

POOH (*abbr.*): Pull out of hole.

pore (*n.*): An opening or space within a rock or mass of rocks, usually small and often filled with some fluid (water, oil, gas, or all three). Also called pore spaces.

pounds per gallon (ppg) (*n.*): A measure of the density of a fluid (such as drilling mud).

power wrench (*n.*): A wrench that is used to make up or break out drill pipe, tubing, or casing on which the torque is provided by air or fluid pressure. Conventional tongs are operated by mechanical pull provided by a jerk line connected to a cathead.

pressure-relief valve (*n.*): A valve that opens at a preset pressure to relieve excessive pressures within a vessel or line. Also called a pop valve, relief valve, safety valve, or safety relief valve.

prevailing wind (*n.*): A wind pattern of the lower troposphere that persists throughout the year, with some seasonal modification

programmable logic controller (PLC) (*n.*): A device used to manage, or control, another device or devices that govern the operation of a system or process. An operator, using an attached computer, can program the controller to maintain a given set of desirable circumstances and to respond to changes or upsets in the system or process by using ladder logic, which is a logic system that operates much like the rungs on a ladder—that is, before the next rung on the ladder can be scaled, the controller must determine that certain conditions are met on the current rung.

pulsation (*n.*): A periodically recurring phenomenon in which something, such as fluid flow or pressure, alternately increases and decreases.

pulsation dampener (*n.*): 1. Any gas- or liquid-charged, chambered device that minimizes periodic increases and decreases in pressure (as from a mud pump). 2. A device used to reduce pressure pulsations in a flowing stream.

pump room (*n.*): 1. On an offshore drilling rig, an enclosed area in which the mud pumps are located. 2. An enclosed area, especially on an offshore drilling rig, in which special pumps, such as ballast pumps, are located.

quick-opening valve (*n.*): A specially designed valve, usually hydraulically operated from a location remote from the valve, which, when actuated immediately, opens the valve.

quick-setting cement (*n.*): A lightweight slurry designed to control lost circulation by setting very quickly.

rack pipe (*v.*): 1. To place pipe withdrawn from the hole on a pipe rack. 2. To stand pipe on the derrick floor when pulling it out of the hole. Also referred to as racking the pipe.

radar (*n.*): Electronic equipment that transmits and receives high-frequency radio waves to detect, locate, and track distant objects. The word is coined from the phrase ra(dio) d(etecting) a(nd) r(anging).

ram (*n.*): The closing and sealing component on a blowout preventer. One of three types— blind, pipe, or shear—may be installed in several preventers mounted in a stack on top of the wellbore. Blind rams, when closed, form a seal on a hole that has no drill pipe in it; pipe rams, when closed, seal around the pipe; shear rams cut through drill pipe and then form a seal. *See* blind ram, pipe ram, and shear ram.

ram blowout preventer (*n.*): A blowout preventer that uses rams to seal off pressure on a hole that is with or without pipe. Also called a ram preventer. *Compare* annular blowout preventer. *See* blowout preventer, ram.

ram bonnet (*n.*): The housing on a ram blowout preventer, inside of which the rams and ram operating parts move when the preventer is operated.

range of stability (*n.*): The maximum angle to which a ship or mobile offshore drilling rig may be inclined and still be returned to its original upright position.

rate of penetration (ROP) (*n.*): A measure of the speed at which the bit drills into formations, usually expressed in ft. (m) per hour or minutes per ft. (m).

rathole (*n.*): A hole of a diameter smaller than the main hole and drilled in the bottom of the main hole. (*v.*): To reduce the size of the wellbore and drill ahead.

ream (*v.*): To enlarge the wellbore by drilling it again with a special bit. Often a rathole is reamed or opened to the same size as the main wellbore.

reamer (*n.*): A tool used in drilling to smooth the wall of a well, enlarge the hole to the specified size, help stabilize the bit, straighten the wellbore if kinks or doglegs are encountered, or drill directionally.

reciprocating pump (*n.*): A pump consisting of a piston that moves back and forth or up and down in a cylinder. The cylinder is equipped with inlet (suction) and outlet (discharge) valves. On the intake stroke the suction valves are opened and fluid is drawn into the cylinder. On the discharge stroke the suction valves close, the discharge valves open, and fluid is forced out of the cylinder.

register ton (RT) (*n.*): A unit of measure of the internal capacity of ships that equals 100 cu. ft. or about 2.8317 m^3.

relief well (*n.*): A well drilled near and deflected into a well that is out of control, making it possible to bring the wild well under control.

remote BOP control panel (*n.*): A device placed on the rig floor (or other locations) that can be operated by the driller to direct pressure to actuating cylinders that turn the control valves on the main BOP control unit.

remote choke panel (*n.*): A set of controls, usually placed on the rig floor, that are manipulated to control the amount of drilling fluid being circulated through the choke manifold. This procedure is necessary when a kick is being circulated out of a well.

remotely operated vehicle (ROV) (*n.*): In offshore operations, an underwater device controlled from a vessel on the water's surface that is used to inspect and operate certain devices on subsea equipment, such as a blowout preventer stack, and that can be used in place of or in conjunction with diving personnel.

reserve buoyancy (*n.*): The buoyancy above the waterline that keeps a floating vessel upright or seaworthy when the vessel is subjected to wind, waves, currents, and other forces of nature, or when the vessel is subjected to accidental flooding.

reserve pit (*n.*): A mud pit in which a supply of drilling fluid is stored.

reserves (*n., pl.*): The unproduced but recoverable oil or gas in a formation that has been proven by production.

reserve tank (*n.*): A special mud tank that holds mud that is not being actively circulated. A reserve tank usually contains a different type of mud from that which the pump is currently circulating. For example, it may store heavy mud for emergency well control operations.

reservoir (*n.*): 1. A subsurface, porous, permeable rock body in which oil or gas (or both) has accumulated. Most reservoir rocks are limestones, dolomites, sandstones, or a combination. The three basic types of hydrocarbon reservoirs are oil, gas, and condensate. An oil reservoir generally contains three fluids—gas, oil, and water—with oil the dominant product. In the typical oil reservoir, these fluids become vertically segregated because of their different densities. Gas, the lightest, occupies the upper part of the reservoir rocks; water, the lower part; and oil, the intermediate section. In addition to its occurrence as a cap or in solution, gas may accumulate independently of the oil; if so, the reservoir is called a gas reservoir. Associated with the gas, in most instances, are salt water and some oil. In a condensate reservoir the hydrocarbons may exist as a gas, but when brought to the surface some of the heavier ones condense to a liquid. 2. A container or vessel that stores fluid, such as a reservoir on an accumulator that holds hydraulic operating fluid.

reverse circulation (*n.*): The course of drilling fluid downward through the annulus and upward through the drill stem, in contrast to normal circulation in which the course is downward through the drill stem and upward through the annulus. Seldom used in open hole, but frequently used in work-over operations. Also referred to as "circulating the short way," since returns from the bottom can be obtained more quickly than in normal circulation.

reverse osmosis (RO) (*n.*): A method of desalting brackish or salt water by passing it through a membrane that is not permeable to salt, and thereby getting fresh water.

rig down (*v.*): To dismantle a drilling rig and auxiliary equipment following the completion of drilling operations. Also called tear down.

rig floor (*n.*): The area immediately around the rotary table and extending to each corner of the derrick—that is, the area immediately above the substructure on which the rotary table and other equipment rest. Also called the derrick floor or drill floor.

righting arm (*n.*): The horizontal distance between the center of gravity and a vertical line through the center of buoyancy of a floating offshore vessel that is displaced from the upright position.

rig manager (*n.*): An employee of a drilling contractor who is in charge of the entire drilling crew and the drilling rig, providing logistic support to the rig crew and liaison with the operating company.

riser adapter (*n.*): A fitting installed on top of the lower marine riser package's (LMRP's) flex joint. It is a fitting to which the first riser joint is attached to the LMRP.

riser angle indicator (*n.*): An acoustic or electronic device used to monitor the angle of the flex joint on a floating offshore drilling rig. A small angle should usually be maintained on the flex joint to minimize drill pipe fatigue and wear and damage to the blowout preventers, and to maximize the ease with which tools may be run.

riser-handling tool (*n.*): A special tool that enables riser joints, as well as the telescopic joint, to be raised and lowered.

riserless drilling (*n.*): An unconventional deepwater offshore drilling technique used on floating drilling rigs. The technique utilizes relatively small-diameter pipe as a mud return line from the seafloor instead of a large-diameter marine riser. Riserless drilling eliminates the need for a large rig that must be able to handle the enormous weights of riser pipe as well as be able to store it. Moreover, drilling without a riser does not require that the rig be able to store and handle the large amounts of mud needed to circulate through the riser. Finally, riserless drilling eliminates the need to set numerous strings of casing to solve the problem of protecting formations whose fracture pressure is close to the pressure required to prevent kicks.

riser pup joint (*n.*): A short length—usually from 5 to 40 ft. (1 to 12.2 m) long—of riser pipe used to space out the riser assembly as it is run to various water depths.

riser spider (*n.*): A device attached to a rig's rotary table that supports riser pipe on a floating offshore drilling vessel.

riser tensioner (*n.*): An assembly of strong cables (lines) connected to the outer barrel of the telescopic joint and a hydropneumatic piston-and-cylinder sheave assembly. The tensioning system supports the weight of the riser joints by applying force (tension) to the outer barrel of the telescopic joint. Tensioner lines are connected to a tensioner ring, or to fixed pad eyes on the outer barrel. The applied tension must remain constant while compensating for vessel heave.

riser tensioner line (*n.*): A cable that supports the marine riser while compensating for vessel movement.

rotary table (*n.*): The principal piece of equipment in the rotary table assembly; a turning device used to impart rotational power to the drill stem while permitting vertical movement of the pipe for rotary drilling. The master bushing fits inside the opening of the rotary table; it turns the kelly bushing, permitting vertical movement of the kelly while the stem is turning.

roustabout (*n.*): 1. A worker on an offshore rig who handles the equipment and supplies that are sent to the rig from the shore base. 2. A worker who assists the foreman in the general work around a producing oil well, usually on the property of the oil company. 3. A helper on a well servicing unit.

ROV intervention system (*n.*): In a subsea blowout prevention system with a marine riser, the equipment that allows a remotely operated vehicle (ROV) to tie into and operate certain functions, such as closing the shear rams and unlocking the lower marine riser package (LMRP) when the normal surface operating system fails. A ported hydraulic stabbing device in the ROV mates with a receptacle in the LMRP's hydraulic circuit and pumps hydraulic fluid into the circuit to operate the selected component.

run casing (*v.*): To lower a string of casing into the hole. Also called to run pipe.

sack (*n.*): A container for cement, bentonite, ilmenite, barite, caustic, etc. Sacks (bags) contain the following amounts: cement, 94 lbs. (42.6 kg) (1 cu. ft.); bentonite, 100 lbs. (45.5 kg); ilmenite, 100 lbs.; and barite, 100 lbs.

sacrificial anode (*n.*): 1. In cathodic protection, anodes made from metals whose galvanic potentials render them anodic to steel in an electrolyte. They are used up, or sacrificed. 2. A positively charged electrode that prevents the electrical current in the water from corroding and deteriorating the structure.

safety of life at sea (SOLAS) rules (*n., pl.*): International maritime rules established to set minimum safety standards for ships and mobile offshore drilling units underway or under tow.

safety stand down (*n.*): A designated time for management officials in a company to meet with frontline workers to discuss safety issues. Usually it is not necessary to cease normal operations during a safety stand-down.

safe working load (*n.*): In crane operations, that portion of a wire rope's nominal strength that can be applied either to move or sustain a load without damaging or breaking the rope. The safe working load of a rope is accurate only when the rope is new and the equipment is in good condition. Because most ropes on an installation quickly become used, the safe working load of a rope is also quickly reduced. For this reason, safe working load is seldom used to denote wire rope strength.

sagging (*n.*): The distortion of the hull of a vessel when the middle is lower than either end because of excessively heavy or unbalanced loads.

samples (*n., pl.*): 1. The well cuttings obtained at designated footage intervals during drilling. From an examination of these cuttings, the geologist determines the type of rock and formations being drilled and estimates oil and gas content. 2. Small quantities of well fluids obtained for analysis.

sandstone (*n.*): A sedimentary rock composed of individual mineral grains of rock fragments between 0.06 and 2 mm (0.002 and 0.079 in.) in diameter and cemented together by silica, calcite, iron oxide, and so forth. Sandstone is commonly porous and permeable, and therefore a likely type of rock in which to find a petroleum reservoir.

sand trap (*n.*): A steel tank placed under the shale shaker into which mud falls after passing through the shale shaker. The shaker removes mainly cuttings from the mud, so solids such as sand and other fine particles fall with the mud into the sand trap. Many of the solids fall out of the mud in the sand trap; those that do not settle out are removed with other specialized solids and control mud treatment equipment, such as desanders and desilters.

saver sub (*n.*): An expendable substitute device made up in the drill stem to absorb much of the wear between the frequently broken joints (such as between the kelly or top drive and the drill pipe).

scale (*n.*): 1. A mineral deposit (e.g., calcium carbonate) that precipitates out of water and adheres to the inside of pipes, heaters, and other equipment. 2. An ordered set of gauge marks together with their defining figures, words, or symbols, with relation to which position of the index is observed when reading an instrument.

scratcher (*n.*): A device that is fastened to the outside of casing to remove mud cake from the wall of a hole to condition the hole for cementing. By rotating or moving the casing string up and down as it is being run into the hole, the scratcher, formed of stiff wire, removes the cake so that the cement can bond solidly to the formation.

screen liner (*n.*): A pipe that is perforated and often arranged with a wire wrapping to act as a sieve to prevent or minimize the entry of sand particles into the wellbore. Also called a screen pipe.

sea anchor (*n.*): An open-ended parachute-shaped device that creates drag but, because water flows through it, slows the speed of a body attached to it.

seep (*n.*): The surface appearance of oil or gas that results naturally when a reservoir rock becomes exposed to the surface, thus allowing oil or gas to flow out of fissures in the rock.

seismic survey (*n.*): An exploration method in which strong, low-frequency sound waves are generated on the surface or in the water to find subsurface rock structures that may contain hydrocarbons. The sound waves travel through the layers of the earth's crust; however, at formation boundaries some of the waves are reflected back to the surface, where sensitive detectors pick them up. Reflections from shallow formations arrive at the surface sooner than reflections from deep formations, and since the reflections are recorded, a record of the depth and configuration of the various formations can be generated. Interpretation of the record can reveal possible hydrocarbon-bearing formations.

self-propelled (*adj.*): 1. Able to move or travel using a built-in power source, rather than requiring an external power source. 2. Of or relating to a rig's ability to move or travel with its own power source.

semisubmersible drilling rig (*n.*): A floating offshore drilling unit that has pontoons and columns that, when flooded, cause the unit to submerge to a predetermined depth. Living quarters, storage space, and so forth are assembled on the deck. Semisubmersible rigs are self-propelled or towed to a drilling site and anchored or dynamically positioned over the site, or both. In shallow water, some semisubmersibles can be ballasted to rest on the seabed. Semisubmersibles are more stable than drillships and ship-shaped barges and are used extensively to drill wildcat wells in rough waters, such as the North Sea. Two types of semisubmersible rigs are the bottle type and the column stabilized.

setback area (*n.*): The location on the rig floor where crew members place stands of drill pipe and drill collars in a vertical position.

settling tank (*n.*): 1. The steel mud tank in which solid material in mud is allowed to settle out by gravity. It is used only in special situations today, since solids-control equipment has superseded such a tank in most cases. Sometimes called a settling pit. 2. A cylindrical vessel on a lease into which produced emulsion is piped and in which water in the emulsion is allowed to settle out of the oil.

sewage treatment plant (*n.*): A system on offshore locations used to render human and other wastes biologically inert before the wastes are discharged overboard.

shaker pit (*n.*): *See* shaker tank.

shaker tank (*n.*): The mud tank adjacent to the shale shaker, usually the first tank into which mud flows after returning from the hole. Also called a shaker pit.

shale (*n.*): A fine-grained sedimentary rock composed mostly of consolidated clay or mud. Shale is the most frequently occurring sedimentary rock.

shale gas (*n.*): Natural gas that occurs in the sedimentary rock called shale. Generally shale's permeability is very low, although its pores may contain natural gas. Consequently, when shale gas is being drilled for, wells are often drilled horizontally and the shale must be hydraulically fractured to provide a pathway for the gas to flow into the well.

shale shaker (*n.*): A vibrating screen used to remove cuttings from the circulating fluid in rotary drilling operations. The size of the openings in the screen should be selected carefully to be the smallest size possible to allow 100% flow of the fluid. Also called a shaker.

shale shaker screen (*n.*): A special wire mesh installed in a shale shaker that allows liquid mud to pass through but traps the cuttings and larger solid particles the mud carries from the hole. Often more than one screen is installed in the shale shaker; also the screens are usually vibrated to help remove the cuttings and solids on top of the screen.

shear ram (*n.*): The component in a blowout preventer that cuts, or shears, through drill pipe and forms a seal against well pressure. Shear rams are used in floating offshore drilling operations to provide a quick method of moving the rig away from the hole when there is no time to trip the drill stem out of the hole.

shear ram preventer (*n.*): A blowout preventer that uses shear rams as closing elements. *See* shear ram.

shoe (*n.*): A device placed at the end of or beneath an object for various purposes (e.g., casing shoe, guide shoe).

shut in (*v.*): 1. To close the valves on a well so that it stops producing. 2. To close in a well in which a kick has occurred.

shut-in (*adj.*); **shut in** (*v.*): shut off to prevent flow. Said of a well, plant, pump, etc. when valves are closed at both inlet and outlet.

sidetrack (*v.*): To use a whipstock, turbo drill, or other mud motor to drill around broken drill pipe or casing that has become lodged permanently in the hole.

sidetracking (*v.*): To drill around broken drill pipe or casing that has become lodged permanently in the hole, using a whipstock, turbo drill, or other mud motor

slick line (*n.*): *See* wire line.

slick riser joint (*n.*): In drilling from deepwater, floating, offshore drilling rigs, a riser joint that does not have buoyant riser modules added to it

slip and cut (*v.*): *See* slip-and-cutoff program.

slip-and-cutoff program (*n.*): A procedure to ensure that the drilling line wears evenly throughout its life. After a specified number of ton-miles (megajoules) of use, the line is slipped—that is, the traveling block is suspended in the derrick or propped on the rig floor so that it cannot move, the deadline anchor bolts are loosened, and the drilling line is spooled onto the draw-works drum. Enough line is slipped to change the major points of wear on the line, such as where it passes through the sheaves. To prevent excess line from accumulating on the draw-works drum, the worn line is cut off and discarded.

slip joint (*n.*): *See* telescopic joint.

slips (*n., pl.*): Wedge-shaped pieces of metal with serrated inserts (dies) or other gripping elements, such as serrated buttons, that suspend the drill pipe or drill collars in the rotary table when it is necessary to disconnect the drill stem from the top-drive unit's driveshaft. Power slips are pneumatically or hydraulically actuated devices that allow the crew to dispense with the manual handling of slips when making a connection. Packers and other downhole equipment are secured in position by slips that engage the pipe by action directed at the surface.

slurry (*n.*): 1. In drilling, a plastic mixture of cement and water that is pumped into a well to harden. There it supports the casing and provides a seal in the wellbore to prevent migration of underground fluids. 2. A mixture in which solids are suspended in a liquid.

SOLAS rules (*n., pl.*): *See* safety of life at sea (SOLAS) rules.

sour crude oil (*n.*): Oil containing a higher-than-usual content of hydrogen sulfide or another acidic gas.

space-out (*n.*): 1. The installation of one or more riser pup joints to obtain the overall required length of the riser from the blowout preventer stack to the telescopic joint. 2. The act of ensuring that a pipe ram preventer will not close on a drill pipe tool joint when the drill stem is stationary. A pup joint may be made up in the drill string to lengthen it sufficiently.

space out (*v.*): To position the correct number of feet (or meters) or joints of pipe from the packer to the surface tree, or from the rig floor to the blowout preventer stack.

space-out joint (*n.*): The joint of drill pipe that is used in hang-off operations so that no tool joint is opposite a set of preventer rams.

spear (*n.*): A fishing tool used to retrieve pipe lost in a well. The spear is lowered down the hole and into the lost pipe. When weight, torque, or both are applied to the string to which the spear is attached, the slips in the spear expand and tightly grip the inside of the wall of the lost pipe. Then the string, spear, and lost pipe are pulled to the surface.

squall line (*n.*): In meteorology, a series of small storms that occur along a cold front.

stability (*n.*): 1. The ability of a ship or mobile offshore drilling rig to return to an upright position when it has rolled to either side because of an external force (such as waves). 2. The ability of a measuring instrument to maintain its accuracy over a long period.

stabilizer (*n.*): 1. A tool placed on a drill collar near the bit that is used, depending on where it is placed, either to maintain a particular hole angle or to change the angle by controlling the location of the contact point between the hole and the collars. 2. A vessel in which hydrocarbon vapors are separated from liquids. 3. A fractionation system that reduces vapor pressure so that the resulting liquid is less volatile.

standpipe (*n.*): A vertical pipe rising along the side of the derrick that joins the discharge line leading from the mud pump to the rotary hose and through which mud is pumped into the hole.

standpipe manifold (*n.*): A part of the mud circulation system that contains several valves, piping, tees, and elbows and is part of the standpipe. Manifold valves direct the mud from the mud-pumping system through the piping and fittings and allow the rig crew to direct the mud to its desired destination, such as to the rotary hose, a backup standpipe, or other piece of circulating equipment.

starboard (*n.*): (nautical) The right side of a vessel (determined by looking toward the bow or forward).

station bill (*n.*): A poster that gives duties and places for each individual on the vessel, rig, or platform for various types of emergencies. Every person on the rig or platform should be familiar with the station bill.

station keeping (*n.*): The means used to maintain a rig or vessel at a particular location. This can be with moorings, dynamic positioning, or possibly tugs, or with a combination.

strap (*v.*): To measure and record the dimensions of oil tanks to prepare tank tables (gauging tables) for accurately determining the volume of oil in a tank at any measured depth. Past tense: strapped.

string (*n.*): An assembly of several individual joints, or lengths, of tubulars, such as drill pipe, drill collars, casing, and tubing, or other lengths of equipment, such as sucker rods. The tubulars or rods are joined to form a continuous length of pipe or rods. For example, connecting several single joints of drill pipe forms a string of drill pipe that can be thousands of feet or meters long; connecting several lengths of sucker rods forms a string of sucker rods that can also be thousands of feet or meters long.

strokes per minute (spm) (*n.*): The number of times all the mud pump's pistons move forward and back per minute.

sub (*n.*): A short, threaded piece of pipe used to adapt parts of the drilling string that cannot otherwise be screwed together because of differences in thread size or design. A sub (i.e., a substitute) may also perform a special function. Lifting subs are used with drill collars to provide a shoulder to fit the drill pipe elevators.

subsea blowout preventer (*n.*): A blowout preventer placed on the seafloor for use by a floating offshore drilling rig.

subsea electronic module (SEM) (*n.*): In drilling from floating drilling rigs in deep water, a device in a multiplex electronic (MUX) control system that contains a programmable logic controller, a modem, and a power supply. An SEM converts signals received from the blowout preventer operating system on the surface to ensure quick operation of the subsea blowout preventer components. *See* multiplex electronic (MUX) control system.

subsea pilot manipulated (SPM) (*adj.*): Of a device that is operated and controlled (manipulated) by a subsea pilot valve. *See* subsea pilot manipulated (SPM) valve.

subsea pilot manipulated (SPM) valve (*n.*): On the control pod of a subsea blowout preventer system, a control device (a valve) employed in the operation of the blowout preventers. An SPM valve is a hydraulically operated, three-position, four-way valve that directs regulated pressure to the blowout preventers.

subsea riser (*n.*): A vertical section of pipe that connects pipeline on the sea bottom to a production platform on the surface. The riser is an integral part of the pipeline and is clamped directly to a leg or brace on the platform.

subsea test tree (*n.*): A device designed to be landed in a subsea wellhead or blowout preventer stack to provide a means of closing in the well on the ocean floor so that a drill stem test of an offshore well can be obtained.

subsea wellhead (*n.*): The equipment installed at the top of the wellbore, which is below the water's surface and on the seabed, and to which the blowout preventer stack is attached.

subsea wellhead and casing system (*n.*): In offshore drilling from floating rigs, the equipment and pipe (casing) that form the foundation of a well and link the wellbore to the subsea blowout preventer stack.

sway (*n.*): The motion of a mobile offshore drilling rig in a linear direction from side to side or perpendicular to a line through the centerline of the rig; especially the side-to-side motion when the rig is moored in a seaway.

swivel (*n.*): 1. A rotary tool that is hung from the hook and the traveling block to suspend and permit free rotation of the drill stem. It also provides a connection for the rotary hose and a passageway for the flow of drilling fluid into the drill stem. 2. A component used to remove rotations of wire or chain (or both). It can be connected directly into chain or wire spelter sockets or used in conjunction with shackles.

synthetic-based mud (SBM) (*n.*): A drilling fluid containing man-made chemicals that emulate natural oil. Natural oil-based muds require that the cuttings made by the bit be specially handled to prevent damage to the environment; synthetic muds were therefore developed to replace oil-based muds in environmentally sensitive areas. For this reason, synthetic-based muds are sometimes termed pseudo-oil-base mud.

tag line (*n.*): In crane operations, a small piece of soft line attached to the bottom of a load suspended by the crane, which, when grasped by a crew member, allows the crew member to prevent rotation of the load.

tank ship (*n.*): A ship designed to transport oil.

tapered hole (*n.*): 1. A condition wherein the hole diameter narrows with depth owing to wear on the bit gauge caused by drilling abrasive formations. 2. A wellbore whose diameter is larger near the surface than near the bottom.

telescopic joint (*n.*): A device used in the marine riser system of a floating drilling rig to compensate for the vertical motion of the rig caused by wind, waves, or weather. It consists of an inner barrel attached beneath the rig floor and an outer barrel attached to the riser pipe and is an integrated part of the riser system. Also called a slip joint.

tensioner system (*n.*): A set of devices installed on a floating offshore drilling rig to maintain constant tension on the riser pipe, despite any vertical motion made by the rig system on a floating offshore drilling rig that uses a hydropneumatic cylinder assembly to which is attached the tensioning lines from the telescopic joint. High-pressure air and hydraulic fluid in the cylinder assembly provide tension on the lines as the rig moves up and down (heaves) with ocean movements.

tension gauge (*n.*): A device that measures and indicates the amount of pulling force (tension) being put on pipe, a tape measure, etc. In tank strapping, the strapping crew attaches a tension gauge to the measuring tape they use to determine a tank's circumference. Knowing how much tension to apply to the tape is essential to accurate measurement, because the same amount of tension must be applied to the tape each time they make a measurement.

tension joint (*n.*): On offshore production systems from floating surface facilities, a riser joint that is positioned directly below the surface joint. It applies additional tension to the riser from within the surface facility's moon pool, using the marine riser-tensioning system.

test plug (*n.*): When a blowout preventer stack is being pressure-tested, a blocking device that is run and landed in the high-pressure wellhead housing. The plug isolates the casing string, which has a lower pressure rating than the blowout preventer stack from blowout preventer test pressure.

thief formation (*n.*): 1. A formation that absorbs drilling fluid as it is circulated in the well. Lost circulation is caused by a thief formation. Also called a thief sand or a thief zone. 2. A low-pressure reservoir or zone that "steals" fluids from a higher-pressure reservoir or zone and prevents their production to the surface.

tight formation (*n.*): A petroleum- or water-bearing formation of relatively low porosity and permeability.

ton-mile (*n.*): The unit of service given by a hoisting line in moving 1 ton of load over a distance of 1 mile. The SI measurement is the megajoule.

tool joint (*n.*): A heavy coupling element for drill pipe made of special-alloy steel. Tool joints have coarse, tapered threads and seating shoulders designed to sustain the weight of the drill stem, withstand the strain of frequent coupling and uncoupling, and provide a leakproof seal. The male section of the joint, or the pin, is attached to one end of a length of drill pipe, and the female section, or the box, is attached to the other end. The tool joint is usually friction-welded to the end of the pipe or screwed on, or both. A hard metal facing is often applied in a band around the outside of the tool joint, to enable it to resist abrasion from the walls of the borehole.

tool pusher (*n.*): An employee of a drilling contractor who is in charge of the entire drilling crew.

top drive (*n.*): A device similar to a power swivel that is used in place of the rotary table to turn the drill stem. It also includes power tongs. Modern top drives combine the elevator, the tongs, the swivel, and the hook. Even though the rotary table assembly is not used to rotate the drill stem and bit, the top-drive system retains it to provide a place to set the slips to suspend the drill stem when drilling stops.

total displacement (*n.*): The lightship displacement plus the maximum-allowed variable load. *See* lightship displacement.

trade wind (*n.*): A prevailing tropical wind blowing toward the equator from the northeast in the Northern Hemisphere, or from the southeast in the Southern Hemisphere.

traveling block (*n.*): An arrangement of pulleys, or sheaves, through which drilling line is reeved and that move up and down in the derrick or mast.

triplex pump (*n.*): A reciprocating pump with three pistons or plungers.

trip out (*v.*): To come out of the hole.

tripping (*n.*): The operation of hoisting the drill stem out of and returning it to the wellbore. *See* make a trip.

tripping out (*v.*): See trip out.

trip tank (*n.*): A small mud tank with a capacity of 10–15 bbls. (1–3 m^3), often with 1 bbl. or H bbl. (decaliter or liter) divisions, used to ascertain the amount of mud necessary to keep the wellbore full with the exact amount of mud that is displaced by drill pipe. When the bit comes out of the hole, a volume of mud equal to that which the drill pipe occupied while in the hole must be pumped into the hole to replace the pipe. When the bit goes back in the hole, the drill pipe displaces a certain amount of mud, and a trip tank can be used again to keep track of this volume.

trip tank indicator (*n.*): A device installed on a trip tank that shows the amount of mud being removed from or added to the trip tank.

true vertical depth (TVD) (*n.*): The depth of a well measured from the surface straight down to the bottom of the well. The true vertical depth of a well may be quite different from its actual measured depth, because wells are very seldom drilled exactly vertical.

tungsten carbide (*n.*): A fine, very hard, gray crystalline powder, a compound of tungsten and carbon. This compound is bonded with cobalt or nickel in cemented carbide compositions and is used for cutting tools, abrasives, and dies.

tungsten carbide bit (*n.*): A type of roller cone bit with inserts made of tungsten carbide. Also called a tungsten carbide insert bit.

turnkey contract (*n.*): A drilling contract that calls for the payment of a stipulated amount to the drilling contractor on completion of the well. In a turnkey contract, the contractor furnishes all material and labor and controls the entire drilling operation, independent of operator supervision. A turnkey contract does not, as a rule, include the completion of a well as a producer.

ullage (*n.*): The amount by which a tank or a vessel is short of being full, especially on ships. Ullage in a tank is necessary to allow space for the expansion of the oil in the tank when the temperature increases. Also called outage.

ultra-deep water (*n.*): Water deeper than 3,000 ft. (914 m).

ultrasonic (UT) inspection (*n.*): An inspection method primarily used for taking thickness readings of steel structures. High-frequency sound waves are introduced to the metal by a diver-held transducer. The sound waves travel through the metal and bounce off the back wall. The instrument measures the time it takes to return to the transducer.

underbalanced (*adj.*): Of or relating to a condition in which pressure in the wellbore is less than the pressure in the formation.

underbalanced drilling (UBD) (*n.*): To carry on drilling operations with a mud whose density is such that it exerts less pressure on the bottom than the pressure in the formation while maintaining a seal (usually with a rotating head) to prevent the well fluids from blowing out under the rig. Drilling under pressure is advantageous in that the rate of penetration is relatively fast; however, the technique requires extreme caution.

underground blowout (*n.*): An uncontrolled flow of gas, salt water, or other fluid out of the wellbore and into another formation that the wellbore has penetrated.

uninterruptible power supply (UPS) (*n.*): An electrical system with built-in redundancy that supplies (powers) an entire electronic control system and provides immediate changeover should one part of the system fail.

unstable equilibrium (*n.*): A condition in which an offshore floating rig continues to incline after an external force caused an initial small angle.

upper string (*n.*): Any part of the drill stem, tubing string, or casing string that lies in the upper part of the borehole.

upstream business unit (*n.*): An oil company or a part of an oil company that explores for, drills for, and produces oil and natural gas. Also known as an exploration and production (E and P) unit.

vacuum (*n.*): 1. A space that is theoretically void of all matter and that exerts zero pressure. 2. A condition that exists in a system when pressure is reduced below atmospheric pressure.

vacuum degasser (*n.*): A device in which entrained gas in the mud returning from the wellbore is removed from the mud by the action of a vacuum inside a tank. The gas-cut mud is pulled into the tank, the gas is removed, and the gas-free mud is discharged back into the mud pits. Also called a centrifugal degasser.

variable-bore ram (VBR) (*n.*): A ram blowout preventer that contains blocks (rams) that can close and seal on a range of pipe sizes—for example, from 3½ to 5 in. (89 to 127 mm) pipe. Variable-bore rams contain a large reserve of rubber in the ram block that specially designed antiextrusion plates force into sealing contact with smaller sizes of pipe. The antiextrusion plates also support the excess rubber when wellbore pressure is applied.

variable-ram blowout preventer (*n.*): A type of ram blowout preventer that has rams that can close on drill pipe of more than one size. For example, the ram preventer can be fitted with a set of variable rams that can properly close not only on 4 in. drill pipe, but also on 4H and 5 in. drill pipe. If a drill string is made up of more than one size of pipe, variable rams eliminate the need for having more than one set of conventional pipe ram preventers available for the pipe sizes being run.

V-**door** (*n.*): An opening at floor level in a side of a derrick used as an entry to bring in drill pipe, casing, and other tools from the pipe rack. The name comes from the fact that on the old standard derrick, the shape of the opening was an inverted V.

vent line (*n.*): An opening in a vessel, line, or pump to permit the escape of air or gas.

vertical center of gravity (VCG) (*n.*): The vertical location about which the weight (mass) of an object is centered. VCG can be determined by multiplying each component weight by its height and then summing those values and dividing by the total weight of the object. VCG is often critical to the stability of floating vessels because excessive VCG can cause a vessel to capsize.

vortex flowmeter (*n.*): In fluid flow measurement, a device in a meter that uses a piezoelectric crystal to measure the rate of flow. As the fluid flows through the meter, vortices (spiral or whirling motions) occur, which the piezoelectric crystal detects. The crystal then produces AC voltage whose frequency is directly proportional to flow rate.

wall cake (*n.*): Also called filter cake or mud cake. *See* mud cake.

wash (*n.*): A thin fluid that separates drilling mud from cement when cement is pumped downhole. A wash also removes mud from the walls of the hole by turbulent and surfactant action.

washout (*n.*): 1. Excessive wellbore enlargement caused by solvent and erosional action of the drilling fluid. 2. A fluid-cut opening caused by fluid leakage.

watch circle (*n.*): Horizontal distances measured from the wellhead (being the center) by the dynamic-position operator to determine at what point an emergency disconnect sequence (EDS) needs to be initiated so as not to damage any subsea or surface equipment. Operators typically have three circles, with the last being red, at which point an EDS is initiated.

waterline (*n.*): 1. The line on the hull of a ship or floating rig to which the surface of the water rises when the ship or rig is floating. 2. Any of several lines parallel to this line, marked on the hull of a ship or rig, indicating the depth to which the ship or rig submerges under various loads.

watertight door (*n.*): A door on ships or mobile offshore rigs that, when closed, blocks the passage of water and withstands its pressure.

watertight integrity (*n.*): On a ship or offshore rig, the vessel's ability to prevent water from entering compartments on the vessel and thus maintain the vessel's ability to remain afloat in spite of several compartments being flooded. To provide watertight integrity, naval architects design vessels so that they are divided into many sections or compartments, each of which can be sealed closed to prevent water from entering—that is, each compartment can be made watertight by closing watertight doors or hatches.

wear bushing (*n.*): A fixed or removable cylindrical metal lining placed inside a fitting or piece of moving equipment that reduces friction and thus wear on the fitting.

wear sleeve (*n.*): 1. A hollow, cylindrical device installed around the swivel's rotating stem that absorbs the stem's rotation and helps form an oil seal between the stem and the oil bath reservoirs inside the swivel. 2. In a riser adapter, a thin metal cylinder that fits inside the adapter and prevents premature wear on the inside bore of the adapter and drill stem components that rotate within it.

weight indicator (*n.*): An instrument near the driller's position on a drilling rig that shows both the weight of the drill stem that is hanging from the hook (hook load) and the weight that is placed on the bit by the drill collars (weight on bit).

weight on bit (WOB) (*n.*): The amount of downward force placed on the bit by the weight of the drill collars.

well completion (*n.*): 1. The activities and methods of preparing a well for the production of oil and gas, or for other purposes, such as injection; the method by which one or more flow paths for hydrocarbons are established between the reservoir and the surface. 2. The system of tubulars, packers, and other tools installed beneath the wellhead in the production casing; that is, the tool assembly that provides the hydrocarbon flow path or paths. *See* completion.

well control (*n.*): The methods used to control a kick and prevent a well from blowing out. Such techniques include (but are not limited to) keeping the borehole completely filled with drilling mud of the proper weight or density during all operations, exercising reasonable care when tripping pipe out of the hole to prevent swabbing, and keeping careful track of the amount of mud put into the hole to replace the volume of pipe removed from the hole during a trip.

well-control equipment (*n.*): An assembly of several components, such as ram preventers, annular preventers, a choke-and-kill system, trip tanks, and mud gas separators. On offshore floating rigs, well-control equipment also includes a marine riser system and a diverter system.

wellhead (*n.*): The equipment installed at the top of the wellbore. A wellhead includes such equipment as the casing head and tubing head. (*adj.*): Pertaining to the wellhead (e.g., wellhead pressure).

well stimulation (*n.*): Any of several operations used to increase the production of a well, such as acidizing or fracturing.

whipstock (*n.*): A long steel casing that uses an inclined plane to cause the bit to deflect from the original borehole at a slight angle. Whipstocks are sometimes used in controlled directional drilling, in straightening crooked boreholes, and in sidetracking to avoid unretrieved fish.

winch (*n.*): 1. A machine that pulls or hoists by winding a cable around a spool. 2. A device used to store, heave in, and pay out mooring wire on a rig. 3. A machine that winds wire rope off and on to a storage drum, using a brake to control the speed of unwinding the wire rope.

wire line (*n.*): a slender, rodlike, or threadlike piece of metal, usually small in diameter, that is used for lowering special tools (such as logging sondes, perforating guns, etc.) into the well. Also called slick line.

work over (*v.*): To perform one or more of a variety of remedial operations on a producing oil well to try to increase production. Examples of workover operations are deepening, plugging back, pulling and resetting liners, and squeeze cementing.

yaw (*n.*): On a mobile offshore drilling rig or ship, the angular motion as the bow or stern moves from side to side.

zerk (*n.*): A special fitting on equipment that accommodates a similar fitting on a grease gun. The zerk allows grease to be injected, but forms a seal to prevent the entry of dirt into the equipment when the gun is removed.

zone (*n.*): A rock stratum that is different from or distinguished from another stratum (e.g., a pay zone).

zone isolation (*n.*):The practice of separating producing formations from one another by means of casing, cement, and packers for the purposes of pressure control and maintenance, as well as the prevention of mixing of fluids from separate formations.

zone of lost circulation (*n.*): A formation that contains holes or cracks large enough to allow cement to flow into the formation instead of up along the annulus outside the casing.

REFERENCES

Baker, Ron. *A Primer of Offshore Operations*. 3rd ed. Austin: University of Texas Press, 1998.

Davis, L. D. *Rotary Drilling: Rotary, Kelly, Swivel, Tongs, and Top Drive*. Austin: University of Texas Press, 1995.

Marcom, Michael R., and K. R. Bork. *The Rotary Rig and Its Components*. Rotary Drilling. Austin: University of Texas Press, 2015.

McCrae, Hugh. *Marine Riser Systems and Subsea Blowout Preventers*. Rotary Drilling. Austin: University of Texas Press, 2003.

Van Dyke, Kate. *Drilling Fluids, Mud Pumps, and Conditioning Equipment*. Rotary Drilling. Austin: University of Texas Press, 1998.

DNVGL-OS-A101: Safety Principles and Arrangement.

DNVGL-OS-D301: Fire Protection.

DNVGL-OS-E101: Drilling Plant.

DNVGL-RU-OU-0101: Offshore Drilling and Support Units.

DNVGL: Rules for Classification of Ships; Dynamic Positioning Systems.

Kinetic Pressure Control Systems. https://shearanything.com/.

Transocean Deepwater Invictus: As-Built drawings; PandID.

Transocean Deepwater Invictus: Dynamic Positioning Operations Manual.

Transocean Deepwater Invictus: Marine Operations Manual.